Determinants of Productivity in Indian Science

Determinants of Productivity in Indian Science

By

A. V. S. Kamesh

CAMBRIDGE SCHOLARS
PUBLISHING

Determinants of Productivity in Indian Science, by A. V. S. Kamesh

This book first published 2010

Cambridge Scholars Publishing

12 Back Chapman Street, Newcastle upon Tyne, NE6 2XX, UK

British Library Cataloguing in Publication Data
A catalogue record for this book is available from the British Library

ISBN (10): 1-4438-2157-8, ISBN (13): 978-1-4438-2157-5

Dedicated to My Family

The good that is there in this, belongs to them by right.

—Graham Greene

Contents

LIST OF FIGURES

LIST OF TABLES

LIST OF TABLES

ACKNOWLEDGEMENTS

"A teacher affects eternity; he can never tell where his influence stops."
—Henry Adams

The influence of my teachers, Professors G. S. Aurora, Ratna Naidu, Vijay Kochar, E. Haribabu, Chandra Sekhar Bhatt, Shasheez, Dr Lakshmi Narayan, Dr Vijaya Tilakam, and Late Dr Abdul Khader Imadi has been overwhelming. I humbly acknowledge it.

I am deeply indebted to Professor Palle Rama Rao, former Vice-Chancellor of the University of Hyderabad, India's visionary of science and technology, for having taken interest in this work and providing constant encouragement.

I thank my parents, brothers Prasad and Bhaskar and Dr Y. V. S. S. Sarma who have been very patient, supportive, and encouraging throughout the course of my work. Bhaskar and Prasad have been friends and critics, providing staunch support along the way. Pavani, Padmaja, little Anav and Aditi also require a special mention.

I am thankful to my colleagues, Dr Kiran Boggavarapu, Dr Murali Krishna Nandigama, Sagar, Dr Pankaz Kumar Sharma, Dr Anoop, Dr Gurrappa, Dr Shree Charan, Vishal, Dr Allen Roy, Dr Shiva, Late Dr G. V. K. S. N. Reddy, Dr Mohan Kumar, Dr Aparna Kumar and Dr Nagalaxmi, for their kind help and timely cooperation. May I use this opportunity to acknowledge their help.

I owe my little exposure to the sociology of science to my classmate Late Mr Sushanth Mund.

I am extremely thankful to Dr Eswarappa Kasi and Dr Nagaraju of the University of Hyderabad who have been with me throughout this work.

The financial assistance from the University Grants Commission (UGC), New Delhi is also gratefully acknowledged.

I acknowledge the support extended by everyone in the Department of Sociology at the University of Hyderabad, India.

—A. V. S. Kamesh

CHAPTER ONE

INTRODUCTION

1.1 General Introduction

Sociology of science is concerned with how and in what ways social factors influence the process of generating knowledge and the product. Over time science has moved from an amateur, self financing activity (as the 18th and 19th centuries) to professional, funded/sponsored, (by public and private sources) activity carried out in big laboratories. In today's world most of the scientific knowledge is produced in organizational settings. Social, cultural, technological and administrative variables related to organizations exercise influence on the process of knowledge production, sometimes, the content of the knowledge is influenced by situational contingencies. So there is a need to understand organizations, which have emerged as the dominant loci of scientific knowledge.

Modern organizations are groups of people coming together for a specific purpose, characterized by own context, socialization, technology and interdependent practices like interactive behavior and co-operation within and with rest of those outside the organization from the point of view of their goals, needs which distinguish them from one another. Organizations with an aim of production recruit people and train them to suit the needs. People join the organizations to achieve their personal needs. The former process called the 'socializing process', and the latter, 'personalizing process' is programmed to blend in such a way that both the entities benefit within a prescribed context, behavior pattern typically known as the organizational culture. Technology comprises an essential component as that of the human resource in an organization. The organizations acquire different tools that aid in production and communication. Different media facilitates the flow of information within and across organizations. Computer mediated communication technology is a new paradigm in which information technology seems to be emerging as an important mode of communication and co-operation in a given social context and across cultures and organizations. The career structures of people recruited in modern organizations are designed in accordance with their productivity, a measure of their performance according to recognized standards. Thus, generally personnel in the organizations are rewarded according to their output.

Various theories have been offered so far to explain the different forms of organizations, the most accepted, though with its lacunae, being the bureaucratic model of Max Weber (1947). Of late, in the field of organizational structure, attention has been turned towards the emergence of 'post-bureaucratic' organizational forms (Drucker, 1988). These new forms are viewed as closely related to the developments in computer-based technologies, in specific Information and Communication Technologies (ICTs). This work appraises the relationships between these new information and communication technologies and its potential implications for productivity among scientists working in different organizational settings in India.

In this work I focus on determinants of knowledge production in Indian science. This work explores implications of socialization into research focusing on the doctoral and post-doctoral training, during which a budding scientist is exposed to knowledge, practices related to community of researchers in a cognitive domain regarding acquisition, exchange and production of knowledge; interaction within research groups and between research groups; information technology as an enabling tool of communication for scientists' productivity and the size of the research work group.

This work examines the influence of information technology on the communication pattern among organic chemists in India. Collaboration is one of the outcomes of communication in science, which in turn influences as patterns and levels of productivity. I try to elucidate the pattern of collaboration among organic chemists by enumerating the factors responsible for collaboration and examine the impact of communication through information technology on the pattern and extent of collaboration.

Keeping in view the fact that productivity is the major parameter of measuring performance in scientific organizations, the focus is on the conditions that facilitate/hinder productivity in the Indian context. Conditions of productivity include opportunities to do research adequate research infrastructure, communication facilities and organizational culture in the Indian context. Sources of ideas for producing knowledge include journals (libraries are repositories of journals), conferences and the recent World Wide Web (www) through which access to journals, the vehicles of communication in science, is made cheaper and faster. Recently e-journals have become an important medium. One of the important facilitating conditions in today's context is the computer-mediated communication. Through computers one can access information available worldwide. Hence, one of the prerequisites is availability of or access to computers. In different organizational settings, the degree of access to computers may vary, especially in developing countries like India. This in turn would influence the extent to which they can gain access to information.

I hypothesize a typical life cycle of scientist in which he starts as a doctoral student, passes through post-doctoral training and is recruited in an organization to start the professional career. Stage-I of dependence with the doctoral supervisor and department, for acquiring knowledge to produce scientific knowledge with the aid of technology related to the process of production, communicating with the peers and networking and being inducted into a regular/permanent job.

The Stage-II of independence starts with the recruitment into a scientific organization, and acquiring the technology in terms of infra-structure for research as well as for communication and building one's own scientific group.

The stage-III describes the interdependence among the hitherto independent scientists for want of information, knowledge and resources in terms of infrastructure for being more productive to achieve organizational mandate or personal goals.

1.2 Review of Literature

Science is a social activity. Its contents-description, explanations and techniques have been created by human beings and shared among groups of human beings engaged in scientific activity. As a social activity, science is clearly a product of history, which takes place in time and place. Science as an act of knowing involves the application of rational, empirical methods and underlying belief in a material reality.

In all modern societies of the contemporary world, science is recognized as a legitimate social activity and various levels of public support are extended to it due to its perceived role in socio-economic and cultural transformation. Comparative analysis of science in different societies would illuminate specific features of the structure and organization of science, values and norms guiding the cognitive activities of the communities of scientists and interaction among themselves at individual and institutional levels and between science on the one hand and economic, political and social power structures on the other. (Haribabu: 1991)

1.2.1 Theoretical Orientations

In the sociological study of science, several theoretical orientations emerged over the years. The first approach is concerned with investigating the interrelationship between science and other social institutions and the influence of these institutions on science and vice versa. The second approach is concerned with the study of science as a social system. Science is treated not as a body of knowledge or as a set of methods and techniques but as: -

> ...the organized social activity of men and women who are concerned with extending man's body of empirical knowledge through the uses of techniques. The relationships among these people, guided by a set of shared norms, constitute the social characteristics of science.
> (Storer 1966:3)

The term scientific community has gained importance in sociology of science. Generally scientific community is defined in a way that a community is defined. Merton (1973) who looked at science as an institution tends to view scientific community as one, whose activities are governed by set of norms. However, Merton makes a distinction between technical norms- logic and evidence- that are employed in evaluating knowledge claims are pre-given and impersonal and hence beyond sociological scrutiny. What a sociologist therefore, can do is to examine as to what extent a scientist's behavior conforms to the social/moral ethos of science. The norms defined by Merton are:

Universalism- openness to scientific knowledge irrespective of nationality, caste, class, gender etc.,
Communalism-sharing scientific knowledge through full open communication,
Disinterestedness-no vested interest on scientific knowledge,
Organized Skepticism- making of a final judgement through methods, not by dogmatism,
Originality-fabricating for the first time and
Humility-learning new things.

A scientist's conformity to norms would be rewarded in the form of recognition and any deviation would attract sanctions/punishments. Further Merton's analysis showed an increasing tendency for the work of the more productive and eminent scientists to receive more and more attention. This latter process, which Merton termed as the 'Mathew Effect' leads to the distribution recognition, reward and communication system to highly recognized/reputed scientists. Merton's functionalist paradigm of the sociology of science has been criticized on the conceptual and methodological grounds, where Mathew Effect within this perspective is unsatisfactory due to its location of "class structure" which places accessing of sources. (Mulkay 1979, 80).

Kuhn's Work

Thomas Kuhn defines scientific community in a more historically more constitutive sense. It is a scientific community shaped by a paradigm, which evolves norms both technical and social. What is technical is socially, historically arrived at. He emphasized on the revolutionary character of scientific progress, where a revolution involves the abandonment of one theoretical structure and its replacement by another, incompatible. He suggested the sociological characteristics of scientific communities.

In *The Structure of Scientific Revolution* (1962), Kuhn says that science never grows by linear growth but by discontinuous revolutionary shift of *paradigms*--a major break through which gives a new research tradition, a whole way of thinking and acting within scientific community. This is because the paradigm represents the totality of the background information, the laws and theories which the members of a particular scientific community adopt for their application.

Normal scientists will articulate and develop the paradigm in response to account for and accommodate the behavior of the some relevant aspects of the real world which revealed through experimentation. In doing so, they will inevitably experience difficulties and encounter falsifications. If difficulties of that kind get out of hand, a crisis state develops. A crisis is resolved when an entirely new paradigm emerges and attracts the allegiance of more and more scientists until eventually the original, problem-ridden paradigms are abandoned. This discontinuous change constitutes *Scientific Revolution* .So, the notion of universality and objectivity is paradigm-bound (Haribabu, 1999).

Post-Kuhnian sociology of science has been attempting not only to understand the organization of science but also the content of science-descriptions, explanations, theories and models in relation to the context (Haribabu, 1999) . Cetina's *The Manufacture of Knowledge* (1981) is a post-Kuhnian development where she introduces *social-constructivist* approach to study the production of scientific knowledge among scientists by employing practical, analogical, literary, symbolical and social situated reasoning. Further, she suggested that *lab* is the site of production such as; Reasoning, Tacit Knowledge and local contingencies.

This production of scientific knowledge is based on local and situational contingencies and influences. In the scientific lab, the goal is not to find the truth but to understand how the works are going on and what is going on. Cetina shows how science draws cultural inputs, resources (physical, cultural) from the wider society. Growth of scientific knowledge is no longer epistemic (working in a particular paradigm) but it works as a transepistemic network for resource mobilization. So, scientists in lab as entrepreneurs who does-investment, returns, and optimization

In recent years, social system perspective of science has gained importance. The changing global context, especially, after the WTO came into existence the norms of WTO impinge on science. Ziman (1996) provided a view of transition from academic science to post academic science elaborating the features of each of them and examines whether the contemporary science is shifting towards a new mode of interactive institutional science. The Mertonian norms of academic science, viz., communalism, universality, disinterestedness, originality and skepticism are institutionalized in a setting where public and private patronage is the principal support of scientific research.

This work employs the notion of scientific community to understand Organic chemists in India. I examine whether there is any stage evolving especially in India, which is characterized by the reshaping of scientific community in the wake of the changing global context, emergence of new, faster, and sophisticated modes of enabling communication technologies and that of network oriented sub-culture of collaboration and their consequences are examined for the productivity, career, and reward structure in a society wherein research is pursued in organizational settings with different levels of resource endowments.

1.2.2 Mode- 2 Production of Knowledge

The norms of post-academic science according to (Gibbons et.al1994) shaped in a post-industrial context in contrast to those of academic science, include,

1. Proprietary knowledge,
2. Local problem solving
3. Authoritarian setting of research agenda
4. Commissioned market-driven support and

5. Expertise oriented around problem solving that is accountable to quality control.

Ziman, Gibbons, Limoges, Nowotny et.al.(1994). While arguing that scientific enterprise is changing propose a theoretical framework to investigate five pervasive charectestics of research. They are:

1. Application
2. Interdisciplinary
3. Networking
4. Internationalization and
5. Concentration of resources.

Their results indicate that research is becoming more interdisciplinary and increasingly conducted in networks at international level. Ziman, in "Prometheus Bound, Science in a Dynamic steady state" hypothesizes that the exponential growth in science seen as a norm by scientists motivated all to publish more papers and resulted in a cumulative pressure on resources in science. This conflict over resources with other priorities led to resources attain a steady state and necessitated adjustments in science as a social system that evolved in conditions of exponential growth. Authors of "The New Production of Knowledge: The Dynamics of Science and Research in Contemporary Societies (1994)", Gibbons et.al. argue that, the internal dynamics of science have generated a new way of producing knowledge. They depict a paradigm shift in performing research in qualitative terms from that of discipline based academic activity to a new research mode called "Mode2" due to the post II World war expansion of the research and education system and entrepreneurial fund raising. They conclude that:

> Internal dynamics are bringing about a transition to different type of science system and throw light upon science as an evolving system of interacting institutions.

1.2.3 Science, Scientists and Developing Countries

Literature on Science in developing countries is scattered through numerous journals, seminar reports and proceedings. But there are only a few empirical studies. Moravcsik (1985) shares the same idea that research on science in developing countries is an unexplored and fruitful area.

A large storehouse of documents and reports on science and technology policies in the developing countries reveal that official documents mainly contain statements of intent and that the existing knowledge of science and scientists in developing countries is incomplete.

On the basis of national statistics compiled by organizations such as UNESCO and OECD (Organization for Economic Co-operation and Development), shortcomings of the developing countries research systems and shortage of available resources have been observed (Rossi 1973). Other authors compared socio-economic conditions with the level of scientific development these countries (Eres 1982). In some of his writings De Solla Price (De Solla Price & Gursay 1975) prescribed quantitative indicators for the developing countries. Research by Garfield (1983) and his Institute for scientific information (ISI) in Philadelphia points out to the low productivity levels and degrees of dependency (articles of developing countries have greater impact when co-authored with scientists from developed countries). (Quoted in Krishna, 1991).

These studies tend to show that none of the developing countries has a 'genuine' scientific community, including India. India has numerically the world's third largest scientific community (Shiva and Bandopadhyaya 1980). An Indian scientist interviewed by Shiva and Bandopadhyay said:

> There is no scientific community in this country.... I meet my Colleagues only abroad ....In a well-knit community, where you are exchanging pre-prints, things are happening and there is excitement here. Our excitement comes by mail from outside. It depends on the postal system. This is the worst part; the spirit is dead (Shiva and Bandopadhyaya, 1980. p.587).

Russell and Galina (1998) while elaborating upon the important differences between the way in which science is practiced in the developing world and the developed world point out that science in developing countries suffers from:

- Isolation from the mainstream of scientific activity,
- The need for the development of an indigenous scientific capacity,
- The lack of critical mass of researchers with respect to most fields of knowledge, and
- The urgency of developing better and more efficient communication channels.

Shrum and Campion (2000) who examined whether, the scientists in the developing countries are isolated; contend that, many scientists are isolated in the sense that they are not part of active research fronts. Indeed variation in levels of contact is one of the key features of scientific and technical organization in developed countries. (Shrum and Mullins 1988). In the developing countries, International and domestic networks are inversely related. The present study focuses upon whether the contact aided by the modern means of communication is emerging as a factor responsible for increase in the productivity levels and visibility by initiating the scientists into these networks.

Turnbull (1997) asserts that science is a set of local practices and each set up works as a knowledge system. Rahman (1975), observes, different national and cultural contexts determine the goals set for basic and applied research. Fields of specialization, social outlook, cultural factors, and philosophies of life influence the conduct of research and the goals set by the scientists. Though it is assumed that the research in developing countries is influenced by the cultural and social environments in which they are immersed, Moravsick (1978) questioned the same and argued that,

> Activities in basic and applied research must by nature adhere to certain universal practices regardless of where, why, how, or by whom they are carried out. Science in DCs is therefore not inherently different from science in the industrialized world, rather it is at an earlier stage of development.

Ignatyev (1989) contends that, Knowledge production of developing countries is limited by transnational factors that shape their education, as well as organizational and legal mechanisms of resource production and allocation. This form of dependence is promoted by various human factors institutionalized in social structure and long-term psychological cultural operations. Developing countries are still dependent on industrialized ones regarding what kind of technology they should buy and from whom, and established mechanisms of professional recognition of scientists and researchers.

Since the knowledge formation process in the developing countries is largely influenced and determined by the western world, a considerable part of their scientific output is foreign to where it is produced. Krishna (1990) negates this hypothesis. For him, the growth of science, organized in terms of specialist groups or small communities sharing a set of 'social and cognitive' values to explore and advance systematic knowledge", is recent historic development. Even though science appeared in its institutionalized form from as early as the late 16th and early 17th centuries, the transformative role of science did not come about until the emergence of professionalized communities in 19th century Europe and 20th century North America. The drive towards professionalization and emergence of scientific communities in 19th century Europe shows that these developments have come about in somewhat 'organic' mould catalyzed by the prevailing political structures and other state-supported structures. Further, sites of production of knowledge such as laboratories had begun to be constituted as part of organizational setting such as universities since the later part of the 19th century. Though Euro centered, the investigation of Ben-David (1971) and others offers that, the social context of the non-western cultures such as India has relevance to the development of institutional factors. Despite the implantation of the modern science from about 18th century onwards, colonial structures separated institutionalization from professionalization. For countries such as India, colonial experience is important considering the social processes of professionalization of science and the emergence of national scientific communities. Colonial science or scientific works in colonial enterprises had little to do with the emergence of the Indian scientific community in its emergent period (1900-1920) (V.V.Krishna 1990). After the 1870s it becomes sociologically meaningful to speak of three categories of scientific personnel, 'gate-keepers' and 'scientific soldiers', who were part and parcel of the colonial scientific enterprises, and native Indian scientists and their missionary supporters who constituted the third group.

The third category of scientists for the first time made organized attempts to undertake basic or fundamental research by the 1920s (Krishna, 1990). Specialist groups, schools, and institutions were constituted in Physics, Chemistry, Mathematics, Biological Sciences and Astronomy by the early decades of this century. By the early 1940s the Indian scientific community made its intellectual presence felt in the international scientific world. There were at least nine fellows of the Royal Society as well as a Nobel laureate in Physics. An Indian scientific community was created which not withstanding its limited spheres of influence, regarded advancing the frontiers of knowledge as a means by which Indian national identity could be established at the international level of science.

Jairath (1984) while discussing the criticisms of the Indian science finds that during the colonial era, science developed in a gradual pace, with minimal state intervention. After independence science developed rapidly with state intervention. This work takes into account the different organizational settings created by the state in which the scientific research is pursued, and attempts to understand the role of acquisition and distribution of enabling technologies coupled with a paradigm shift in terms of media of communication and resultant collaborative strategies on the productivity of scientists.

1.3 Scientific Communities and Growth of Knowledge

Habermas (1984) differentiated three primary generic cognitive areas in which human interest generates knowledge. These areas determine categories relevant to what we interpret as knowledge. That is, they are termed 'knowledge constitutive'- they determine the mode of discovering knowledge and whether knowledge claims can be warranted. These areas define cognitive interests or learning domains, and are grounded in different aspects of social existence-work, interaction and power.

Habermas broadly refers to work as the way one controls and manipulates one's surrounding environment, which is known as instrumental action- knowledge based upon empirical investigation and governed by technical rules. The criterion of effective control of reality directs what is or not appropriate action. Habermas classified the 'scientific' research domains-e.g., Physics, Chemistry and Biology which are empirical-analytic using hypotheco-dedcutive theories. We consider scientific work related to research in this framework and examine the production of knowledge in science. In our study we shall examine how the Organic Chemists in the Indian context utilize the organizational context in which they are located in pursuit of being productive.

(Price (1963) advocates that, in science, the certification of knowledge claims leads to addition to or the growth of a cognitive field through a series of stages such as:

(1) A preliminary period of growth in which the absolute increments are small although the rate of increase is large but steadily decreases

(2) A period of exponential growth when the number of publications in a field doubles at regular intervals as a result of a constant rate of growth that produces increasing amounts of absolute growth.

(3) A period when the rate of growth declines but the annual increments remain approximately constant; and

(4) A final period when both the rate of increase and the absolute increase decline and eventually approach zero.

Price (1963) argued that basic science is currently in the second phase of growth but that, as a result of shortages of resources and manpower, it will eventually enter the third an fourth stages. Price does not offer an explanation for such a growth of knowledge. Diane Crane (1972) offers the thesis that the logistic growth of scientific knowledge is the result of exploitation of intellectual innovations by a particular type of scientific community. Productivity of scientists also depends on what stage of the cognitive development of a research field that a scientist enters. She examines the issue as to how scientific communities affect the growth of knowledge. The author comprehends complex sociological problems as that of understanding the social institutions that produce ideas. In dealing with these types of phenomena, the sociologists face the problem of understanding the ideas that can be highly technical and obtuse. Diana Crane (1972) argues that scientific communities are distinct entities, interconnected with one another in ways that are as yet only vaguely understood.

Haribabu(1999) provides a sociological review of the production of knowledge including the social origins of knowledge. The author says that, the rationalist-positivist epistemology of science influenced the streams of thought in sociology of science, more prominently the functionalist paradigm of Robert Merton (1973). Through his empirical studies, Merton analyzed that the inequality in science due to skewed distribution of recognition and rewards is functional in the sense that the recognized scientists provide role model for young scientists as significant others.

Mulkay (1979, 80) criticized Merton on theoretical and methodological grounds for a-historical characterization of science.

Kuhn (1970) a historian of science proposed the concept of paradigm that creates conditions for the emergence of a scientific community with shared cognitive beliefs.

Kuhn defines scientific community in more sociological terms. A paradigm guided scientific community, possesses a shared cognitive belief system, standards of research, communication among the members of the scientific community becomes more focused. After the acquisition of paradigm in any given specialty there will be cumulative growth of science, which Kuhn calls the normal science phase. As the paradigm gets extended to more and more areas of research, it is at this stage that the anomalies emerge, which leads to crisis, a new paradigm emerges, as a response to crisis. The productivity of the scientists who enter the cumulative phase of the growth of the paradigm will be high. During the period, when paradigm faces anomaly and crisis, the productivity of scientists declines. This decline of the paradigm is related to exhaustion of paradigm. Thus scientific knowledge grows according to the cognitive consensus among the members of the scientific community insulated from the society.

Post-kuhnian sociology of science challenged the kuhnian notion of scientific community being insulated from the wider context. Approaches like the social-constructivist approach Knorr-cetina (1981) to the study of production of knowledge focused upon the internal and external social context in which science is situated. Knorr et al (1978) begin with the idea that scientists may be affected by organizational settings and structures than is usually recognized in stratification studies. They examined the effect of organizational factors on productivity and concluded that the rate of publication of research findings by the academic scientists increased steadily on average through out the first twenty years of research and then continued at a fairly highly for the remainder of the career. This increase in research productivity occurred despite the fact that less and less time was spent on research and more time on teaching and administration.

The crucial feature explaining scientists' varying rates of productivity was, whether or not they occupied a supervisory position; The significance of supervisory positions was closely associated with the control they have over manpower resources, physical and human and the extent to which they enabled supervisors to extend the range of research projects in which they were involved. Supervisory scientists were able to maintain increasingly high rates of authorship (or co-authorship) by participating in the research project in a rather different manner from their subordinates. Increasingly they confine themselves to defining projects, which were to be carried out in detail by graduate students and other less senior researchers. To quote Knorr-Cetina (1978) "the higher a scientist moves up the hierarchy of the research laboratory, the more he is confronted with a variety of scientific and nonscientific functions in addition to research; the more his research activities change towards goal setting rather than goal executing; and the more he is able to attract project money and consequently become involved in greater number of research projects."

Thus, it seems that structural factors within the organization impinge on the productivity of researchers, to some extent apart from individual differences in ability and motivation. For high levels of formal productivity are closely associated with the position of the supervisor the availability of such position will depend to a considerable degree upon the institutional system within which it is embedded. Related to the ability of scientists to mobilize funds for research is their ability to constitute research groups.

Similarly micro-level ethnographic studies Latour (1979) revealed that in the production of knowledge social and cultural factors shape the social action of the scientists.

In this work I examine how the context of the organization shapes the action of a scientist within and outside the organization in which he/she is located. We considered that it is the organizations that provide the context for productivity of the scientists with human, technological resources like research groups, peer groups.

I considered the impact of the doctoral institution and location and duration of postdoctoral training in a particular organization on the recruitment, career and later productivity levels communication facilities and opportunity to collaborate. We also attempted to study the impact of the relative access of communication technology and collaborative nature of the scientists on their productivity levels.

1.4 Perspectives on Productivity

In all the organizations including the scientific organizations the work output of the members of the organization is perceived as their productivity. Especially in the scientific organizations, scholars have developed various measures of indicating the productivity of the scientists which include number of publications and citations. "The social inequality in science has been explained in terms of differential productivity, rewards, and recognition (Mulkay 1981)". In this work, we considered productivity as the number of publications by the scientists in refereed journals in a specific time period. In a critical review of the publication productivity of the scientists, Mary Frank Fox (1983) argues that publication is central to research and offers a review of the explanation of productivity levels among scientists through psychological traits, work habits, demographic characteristics and environmental location. Merton (1973) accounts for productivity differentials in terms of only individual talents. Diana Crane (1972) offers the thesis that the productivity of scientists depends on what stage of the cognitive development of a research field that a scientist enters. Long, J. Scott and Robert Mc Ginnis. (1981) describe the impact of the organizational context on the scientific productivity'

Literature shows that,

a) Small proportion of scientists produces the bulk of the research literature.

b) The honorific rewards of science are a monopoly of relatively small intellectual elite. Recognition of scientist's knowledge is expressed in a great variety of ways, from eponymy and the Nobel Prize, through a multiplicity of other prizes and medals, to membership of national academies.

One prominent feature of social ranking in science is that a very few universities, departments, grand ecoles and research organizations are widely regarded as the main centers of excellence. The attainment of elite status is closely bound up with this phenomenon. These main centers of excellence are endowed with the state of the art technology and recruit the best of the scientists. The most eminent scientists end up at an elite institution and that a large proportion also begin their careers within them. Empirical studies reveal that rates of social interaction within the elite tend to be higher than those between elite and non-elite scientists. Most prestigious universities, and the leading scientists within them, receive a disproportionately large share of research funds. This is very much prevalent in developing countries like India, as the National research communities are themselves stratified in terms of productivity; the receipt of major awards and the resources devoted to research. These communities appear to comprise of haves and have-nots in terms of knowledge, access to technology, especially information and communication technology. We observe:

a) Inequality in rates of productivity
b) In access to research facilities

To begin with the productivity,

1. There is a strong empirical connection between research productivity and scientific rank (Gaston)
2. The production of information tends to precede the receipt of rewards
3. Any complete theory of stratification in science will have to cover productivity and resources, as well as rewards, within a single overall account.
4. The notion of universalism as formulated by Merton involved to main elements
    a) The extent to which performance is accessed by means of impersonal criteria
    b) And the degree to which individuals have equal opportunities to perform
5. The productivity of the members of each generation of research scientist becomes increasingly unequal as their careers unfold.

Gaston (1970) summarized that those scientists who manage to establish a comparatively high rate of productivity at the beginning of their career are able to use the initial advantage to claim more research facilities and thereby to improve their relative position further. Thus, Merton (1973) accounts for productivity differentials in terms of only individual talents.

The literature surveyed showed that productivity differentials are socially produced by factors such as:

- Organizational context
- Stage of Development of a Research Field
- Communication and
- Collaboration

The functional analysis of social stratification in science by Cole and Cole reveal that the stratification system serves three major functions.

1. The self-reinforcing process of cumulative advantage helps to ensure that research resources go disproportionately to those scientists who have the greatest ability.
2. The members of the scientific elite act as models for their colleagues and particularly for the younger men. They display clearly their rewards which accrue to scientific excellence and which need to be visible if they are to operate as effective incentives. The existence of distinct elite encourages other scientists to strive for eminence themselves and in doing so to emulate the behavior of those who most embody the values of science.
3. If they were no elite fewer bright students would receive best available training. As long as the elite exists and is located in a few universities of high repute, the most promising student can choose – or be recruited into those departments where they native abilities will be further nurtured through close contact with the best minds of preceding generations.

Cole and Cole (1973) defined the quality of research papers operationally as the number of citations they receive. They argue that when variations in quality are statistically controlled other sociological variables such as rank of Doctoral department, Gender and Membership of an elite university have very little impact on scientist's receipt of rewards. They note that citations can be used not only as a measure of quality, but also as a measure of recognition.

If they were no differentials in distribution of recognition research resources would be more evenly distributed. Some scientists would surely see this as a positive good. But from the point of view a system it is probably advantageous to have the bulk of research resources go to the men who have been successful in the past and to give them the resources to make additional discoveries. There are three major kinds of inequalities in science.

One of the important aspects of stratification of science, missed by the functional analysis is that, rewards, once unequally distributed can be used by those at the top as resources, not only for further increasing their own productivity and the significance of their work, but also, for regulating others research opportunities. From the perspective of the functional theory, differences in rank between individuals within fully institutionalized national research communities are thought to be mainly due to marked differences in individual ability which in a system with minimal structural constraints and objective criteria of research performance, led fairly, smoothly to the emergence of a highly productive and highly rewarded elite.

The second line of analysis, however, depends on the assumption that differences between countries in scientific productivity are themselves socially produced; for instance, By variations in the availability of training facilities, Research resources, Cultural capital and Informal access to information about current developments at the frontier of the knowledge.

1.5 Organizational Context and Scientific Productivity

The organizational context assumes an important role in determining the productivity of the actors. In their study of organizational context and scientific productivity, on bio-chemists, Scott Long and McGinnis (1981), quote Krohn (1971) that organizations are endowed with structural facilities and the context creates the social environment for scientists to perform their work. Organizational context is also important from the point of view of the training it bestows upon the members. Thus organizations shape their research and people who do it. The authors explain that the context determines the research setting and technology associated with it and also affects professional prestige. The organizational context provides an ambience of intellectual autonomy and opportunity for research in terms of freedom to publish. The authors also relate the organizational context to scientific productivity by saying that different goals of various locations may either hinder or promote productivity.

In this work I examine the importance of the context of the organization on the later productivity of scientists by attempting a sociological analysis of the factors like age, duration of association with organization of current affiliation, academic training, similarity of the research area, communication technology and pattern of collaboration. In our study we hypothesized that with reference to Indian context, postdoctoral training shall have significance to the productivity and career of a scientist.

1.5.1 Age and Production of Knowledge

In the realm of science, the production of knowledge is effected by a prolonged period devoted to acquiring knowledge, skills, attitude, values and behavior patterns of the scientists i.e., socialization required to qualify for scientific work. Age and productivity relationship has been explained by sociologists by emphasizing on the reward structure of science. (Hagstrom 1965; Merton 1957, 1968; Reskin1979). The sociological explanation is of the view that age reflects an unfolding career as the older the individual the more history the individual accumulates. Merton (1973) claims that age and production of knowledge are related in scientific research. The production of knowledge is effected by a prolonged period devoted to acquiring knowledge, skills, attitude, values and behavioral patterns of the scientists. i.e., socialization required to qualify for scientific work. Age structure effects both the normative system of science and its intellectual development and reflects the counteracting pattern of life course and cohort flow. The changing age and educational structure of science is congruent with the self-image, prevalent among scientists, that pictures science as a young man's game. Age structure also indicates the time of life at which scientists do their most important work or the relation between age and scientific productivity.

Merton proposes that, during the professional career, scientists perform multiple roles like, that of the roles of research teaching administration and gatekeepers. I argue that at every stage, and in every role scientists perform to add to the production of knowledge. According to Merton, the role of researcher is most important for the scientists.

Lehman(1953) claimed that older scientific community is less productive than a younger one. Bayer and Dutton (1977) in their study on the earth scientists claim that highest peak in productivity comes in the first ten years and the second peak when close to the retirement. Pelz and Andrews (1966) describe that productivity of the scientists reaches two peaks in their career. The first peak occurs during the late 30s and early 40s and the second peak around the age of 50 years. Cole (1979) mentions that productivity in terms of article counts increases up to the age 40-45 but begins to drop off by the age of 50. In this work I attempted to examine what is the most productive phase of one's own age among the organic chemists in India?

1.5.2 Prestige of the Doctoral Department

In their study of Organizational context and scientific productivity, on bio-chemists, Scott Long and McGinnis (1981) found that, chances of obtaining employment in a particular context are not strongly affected by productivity. They found that, scientists in prestigious institutions increased their current productivity independently of earlier productivity. They argue that scientists are not allocated to various contexts on the basis of contributions to the body of scientific knowledge. They observed that, recruitment operates independently of productivity and claim that the stratification system in science is far from universalistic and initial advantages when accumulated, result in the inequality among the scientists in the levels of productivity.

1.5.3 Post Doctoral Training

> "It is largely these postdocs who carry out the sometimes exhilarating, sometimes tedious, day-to-day work of research. Many of them will go on to uncover fundamental new knowledge, chair prestigious academic departments, and form the fast-growing technology companies that power our economy, It is largely they who account for the extraordinary productivity of science and engineering research in United States" (Singer, Maxine, 2000)

Postdoctoral training has become a significant aspect in scientific education and professional cycle of scientists. Paul.D.Allison and J.Scott Long (1982) mention that postdoctoral training is important for public policy making and approvingly quote Reskin (1976) who argued that postdoctoral training position provides a site for understanding stratification and productivity in science. According to the authors, postdoctoral positions besides conferring prestige have important career benefits by adding value to a scientist's work and resulting extra research experience. Postdoctoral training provides young scientist with additional time to demonstrate his or her competence through additional publications. The organization context in which a scientist receives postdoctoral training has more influence in comparison to the influence of doctoral department on later career opportunities. Examination of the association between postdoctoral training and later publication pattern helps answering queries about effects of organizational context on scientific productivity.

According to the National Research Council of United States of America (1976), a postdoctoral appointment is,

> …a temporary appointment, the primary purpose of which is to provide for continued education or experience in research, usually, though not necessarily under the supervision of senior mentor. Included are the appointments in government and industrial laboratories, which resemble in their character and objectives postdoctoral appointments in universities.

Coggshell et.al (1978) in their study on biomedical scientists mention that published research by postdoctoral fellows acts as a credential to permanent employment positions. Scott Long, Allison and McGinnis(1979) in their work on the biochemists' entry to the academic career suggest that. William Zumeta(1984) argues that postdoctoral fellows have become a " key factor of production" in academic research and postdoctoral training is associated with the increase in the levels of research productivity. In the Indian context, we observed that postdoctoral training in scientific research occurs in two contexts- One in India and the other in institutions abroad. Muniyappa(2007) mentions minimum of 5 years of duration of postdoctoral position in foreign countries as a prerequisite for academic career. In this work I examine the impact of the duration of postdoctoral training both in India and abroad on the productivity levels of the organic chemists to understand whether the former provides a site for understanding stratification and productivity (Reskin 1976) in Indian Science.

I attempt to understand the impact of the organizational context in which an Indian holding a doctorate degree in Organic Chemistry receives postdoctoral training, on the productivity later in career.

1.5.4 Stage of Development of a Research Area

The structure of research groups and formation of scientific careers seem to be linked to the growth and decline of research fields. Entry at an opportune or in-opportune moment is a crucial influence on participants' career. Early access to emerging problem areas and access to research resources are important influences, within research laboratory on productivity and careers; and these factors probably play an important in generating process of cumulative advantage. They are equally significant at the institutional level. The members of the relatively few centers of scientific excellence have more resources than other scientists, and these resources are to some extent available to their students and protégées, to assist the latter in achieving high levels of productivity. In addition, eminent scientists tend to be the focus of numerous informal communication networks. They are accordingly well informed about recent developments and are often able to guide their students and junior colleagues into a fruitful or fashionable research area during its initial stage of development when the chances of making a significant contribution are relatively high. Besides efficient selection procedures ensure that most potentially gifted researches come under the tutelage of the eminent at the beginning of their careers.

1.5.5 Technology and Production of Scientific Knowledge

Technology is one of the essential factors for production of knowledge in science. Two kinds of technology are employed in scientific knowledge production. a) Infrastructure related technology: equipment, gadgets, and measuring instruments and b) Communication related technology-Telephone, Fax, E-mail and Internet etc., The two kinds of technology are considered as enabling technologies in a research laboratory and the laboratories are techno-social structures as the interaction within a research group is mediated by technology.

Information technology seems to result in the emergence of a new era, which involves technological and informational reality as a new social reality. We shall examine communication-related technology in detail as we, in the Indian context examine the impact of information technology on communication behaviors and patterns of communication among scientists. Our study aims at understanding the social consequences of the role of technology for organizations and their structures which are under transition from an era of hard-copy based communication to an era of soft-copy based communication.

The effects of new technologies like that of the information technology introduced in the realm of science in developing countries with differential distribution, promise consequences for science and scientific community in terms of acquisition, production of knowledge and productivity of the scientists. Such an assumption leads one to analyze the pivotal relationship between the communication technology and organizational context. Two perspectives may be useful. They are:

- Technological Determinism
- Social Interactionism

Technological determinism is a utopian view of technological innovations that emphasizes their inherent technical worth. It views organizations as machines; highly ordered and efficient work environments dominated by clearly defined rules that all staff members are aware of and generally accept. Ideally shared goals, consensus, and stability typify the technological determinist organization. The technological focus of this perspective ignores the particularities of a new technology's organizational context as it accepts Morgan's (1986) metaphor of organizations as machines.

The social interactionist approach to technological implementation arises from the analyses of how individual organizations work in practice rather than how they ought to work. (Kling 1980, Markus and Robey 1988) say that cause and effect relationships in the implementation of computer-based systems are often difficult. Implementation is thus revealed to be a process of social interaction between the technology and a particular organizational context. According this perspective, technology is not only equipment and sets of methods, but also individual perceptions and understanding of the role and value of such facilities (Sproull and Goodman1990). Individual perceptions of the nature and utility of technology are likely to vary considerably according to personal, organizational, and cultural circumstances. Hence technologies, such as computer-mediated-communication, are not seen as neutral configurations of equipment; rather, they are socially constructed within each organizational context and likely to be continuously reinvented by their users. (Rogers 1993). An important element of the social interactionist perspective is its view of organizations as unique social systems or cultures that socialize individuals into a particular set of norms, beliefs and values. The key assumption of the social interactionist perspective is that technologies do not function independently of their environments; rather they gain meaning only as individual staff members in a particular cultural and organizational context interact with them.

1.6 Communication in Science

Communication is the exchange of information between individuals by means of a common signal system. Scientific communication can be defined as the combined processes of presentation, delivery and scientific information in human society (Mikhailov, 1984,p.39). These processes form the basic mechanisms for the existence and development of science. The object of scientific communication is the registration, evaluation, dissemination and advancement of human knowledge.

Development of science is dependent on production of systematic body of knowledge that characterizes all scientific research. The production and certification of this knowledge involves unlimited interactive behavior among the scientists, research groups, funding agencies. Thus communication and flow of information establish as essential components of scientific research. In the year 1979, William D.Garvey made the famous claim that. ...Communication is the essence of science. (Blau: 1974:39) claimed that the aim of all scientists is to contribute to the body of knowledge in their field and their ability to contribute to the advancement of science depends to an extent upon the amount and the quality of their interaction with fellow scientists.

Communication among scientists is widely recognized as the key to the development of ideas and to the success of individual scientists who work with those ideas. Journals are vehicles of *communication* among members of a scientific community. Evaluation of manuscripts submitted for publication is institutionalized through referee system. Referees are status judges charged with the responsibility of evaluating the quality of role performance of scientists (Zuckerman and Merton 1973:460). Communication as a social process in the realm of production of scientific knowledge plays an instrumental role in increasing the productivity and visibility of scientists. Communication in the realm of science occurs in two forms, formal and informal. Formal communication occurs through publications, and communication events like seminars, conferences and workshops wherein ideas are exchanged. An 'invisible college', traditionally, in a research specialization mediates informal communication. This is a small group of researchers that regularly exchange information about the progress on the 'research front' (Crane, 1972, Lievrouw, 1990).

Kronick (1988), argued that scientific research is essentially a corporate activity and a distinctive feature and accepted social norm of the scientific community is the communality of ideas, a shared commodity that belongs to every one. Thus various studies indicated that scientists expend large proportion of their time on information and communication activities than most other workers.

Menzel 1966 (1000-1003) isolated ***five characteristics*** that conceptualize communication process. They are:

1. Scientific communication is a system, in which the constitutive acts, including the methods by means of which scientific messages are transmitted. Thus mentions about scientists as generators and users of information that constitute interconnected publics.
2. Effective transmission of messages to recipients involves many synergistic channels of communication.
3. Informal communication plays vital role in science communication system and characteristically has regularity in its pattern.
4. A number of communication interests and norms of behavior link scientists and they constitute publics.

Communication system in science serves many functions.

Garvey and Griffith (1968:129-130) proposed the below mentioned ***seven functions*** indicative of scientific communication.

"Answering specific questions, being aware of the contemporary developments in a particular field, understanding a new research paradigm grasp major and emerging trends in a specific field and relative importance of that field within its broader discipline, verification of information reliability by offering additional information, reorient or expanding one's own interest field and Receiving a constructively critical feed back to researcher's own work.

Leah A.Lievrouw and Kathleen Carley (1990) defined:

> ***Communication process***: as any activity or behavior that facilitates the construction and sharing of meaning among individuals, that they consider to be the most useful or appropriate in a given situation. For example, the presentations of research findings in a journal as a publication.
>
> ***Communication Structure***: as a set of relationships among individuals who are linked by the meanings they construct and share. For ex, the research front, invisible college.

The authors elaborated upon the term "Telescience" defined as the existence of geographically dispersed intensely communicative research groups and collaborators, electronic journals, and teleconferences. These mechanisms seem to involve computers to a maximum extent and they gave rise to the idea of computer mediated communication technology (CMC). Telescience adds to the idea that communication channels like e-mail, voice mail, and fax increasingly aided by CMC have the potential to have implications to the stratification system in science, especially in the Indian context. In other words, the relation between level of technology, social contexts and career aspirations of the scientists through reward system is comprehended.

The onset of communication technology viz., Telescience has brought about a revolutionary change in communication pattern among scientists against the scale of time, contributing to a new pattern of extended collaboration. This in turn has an effect on the productivity rates and collaborative patterns and consequently, the system of rewards. Kircz (1998) sought to analyze traditional scientific communication in the wake of the application of electronic revolution in scientific research. He listed four important functions in scientific communication:

1. The certification function- concerns the validation of research quality and is associated with scientific standards within a research program.
2. The-registration function- relates to particular research to individual scientists, who claim priority for the research. This function is closely related to ownership protection, and the reward system.
3. The awareness function- that leads to information needs and the resultant exchange process within the scientific community.
4. The archival- related to storage and accessibility of information.

1.6.1 Communication and Production of Knowledge

Diana Crane (1972) elaborated upon the role of communication in the production of knowledge. Crane argues that logistic growth of scientific knowledge is a result of exploitation of intellectual innovations by a particular type of scientific community.

Crane propounds that the growth of knowledge in a discipline is indicated by sets of clusters of research papers in research areas of the discipline. These research areas are not separate entities, but social and ideational links hold them together permitting diffusion of ideas from one area to another. Thus the growth of research area is indicated by intense communication amongst the members of the community.

Crane suggests that ideas diffuse more effectively when transmitted by individuals rather than by publications, emphasizing more on the informal aspect of communication in science. The author further indicates that scientists in a research area organize themselves as two kinds of groups. The first is that of collaborators and the second group is the network formed by the collaborators with constant interactions among themselves, namely an ‘invisible college’.

Thus scientists discuss ongoing research with other scientists in order to obtain advice and also information about similar studies. Some other with other scientists and publish results joints. This represents a very intense form of communication.

This work explores the sources and diffusion of ideas, which facilitate interaction among researchers within and among research area, especially,, the role of computer mediated communication in scientific work.

1.6.2 The Social Context

Recent research in the sociology of innovation according to Walsh and Bayma (1998) has been driven by the question of the relationship between social structure and technology. Walsh and Bayma (1996a) find that different fields used CMC technology very differently, and that the work organization of a field (Whitley, 1984) seemed to be associated with the extent of and types of use. This is of much significance to the present study, as the comprising units have unequal access to CMC technology. CMC may be particularly suited for coordination, especially among geographically dispersed work groups. Empirical studies by Walsh and Bayma (1996b) reveal that computer mediated communication technology (CMC) facilitated larger collaborations and had consistent relationship with productivity of scientific groups.

1.7 Collaboration in Science

Science is a social activity with practitioners of science performing their research activity in cooperation and coordination within and among the research groups. The advancement of science requires shared knowledge. Presentation of research findings is an integral part of scientific communication. Collaboration among researchers is also an important part of maturation of science. Collaboration is a consequence of communication. Collaboration may be defined as "production interaction between two or more individuals with a conscious goal of producing a concrete output. (Haribabu 2002)

The changing landscape of collaboration in modern science is characterized by the synchronous and asynchronous communication across cultures (Walsh and Maloney 2002) and the emergence of collaboratories wherein collaboration takes place among different laboratories with a proper division of labor in the collaborative project effort. Collaboration may be defined as "production interaction between two or more individuals with a conscious goal of producing a concrete output. (Haribabu, 2002). Collaboration leads to increase in the number of publications. Productivity and Collaboration are correlated (Price 1986). Collaboration is a norm in modern science (Bever and Rosen, 1979b). Several other scholars pointed out that collaboration increases research productivity (Lotka,1926;Price &Beaver,1966; Zuckerman,1967;Godin and Gingras,2000). Sooho Lee and Barry Bozeman(2005) provide with model of relationship between collaboration and productivity. The model

depicts relation between individual, institutional and environmental factors on the one hand and collaboration and productivity on the other.
There are several incentives to collaborate.

1. As scientific research becomes more complex, the expertise and talents of others may provide useful assistance. The lone scientist may be unable to complete many projects. Collaboration fills demands for advanced quantitative and formal techniques, data collection and analysis, and grants contracts.
2. Students and junior colleagues may learn more from active participation in research with skilled senior scholars. Conversely, scholars at later stages of their careers may benefit from new ideas and fresh approaches.
3. Collaboration leads to increased efficiency in output. Scholars acting in concert (through mutual support or through division of labor) may be able to produce greater output than the total of each individual acting independently.
4. The quality of research can be improved with joint participation. The strengths of two or more cooperative researchers can be fused to produce improved, holistic results.
5. Scientific research is a social process. Just as researchers build on past works of others, knowledge increases by direct interactions among scientific researchers.

Reviewing the history of scientific collaboration in the post-World War era, Price (1986) claimed that research grants lead to greater collaboration. He found that a "developing entrepreneurial tradition of channeling research support funding through a principal investigator in effect permits that person to purchase subsidiary authors. The result is that the number of authors per paper has become a rather good indicator of the extent of grant support in the field" (1986:259). Price also noted that collaborative writing might be influenced by the system of scientific communication. When scientists are distant, they make contributions through written communication; as they come together socially or technologically, they collaborate in research endeavors.

1.7.1 Aspects of Collaboration

Successful collaboration involves clear expectations and equitable division of responsibilities and collaboration leads to increase in the number of publications. Productivity and collaboration are correlated. (Price, 1986)

Various studies have emphasized different determinants of scientific collaboration. Beaver and Rosen concluded that scientific collaboration represented a response to increasing professionalization of science (Beaver and Rosen 1979). Collaboration provided a means to both professional advancement and increased knowledge. It offered access to resources (both information and equipment) and association with scientific elite, which enhanced the visibility of aspiring young scientists in particular. (Beaver and Rosen 1979). Hence Beaver and Rosen explained collaboration by the intellectual material sources.

Price emphasized the importance of economic factors in collaboration. In his essay, he reflected that collaborative authorship 'arises more from economic than from intellectual dependence and that the effect is often that of squeezing full papers out of people who only have fractional papers in them at that particular time. (Price 1986, 190)." He went on to assert "the amount of collaborative authorship measures no more than economic value accorded to each field by society (Price, 1986, 160). Price referred to amount of money society allocates to research. We could think of other meanings of economy, such as the advantages of sharing the resources to buy and maintain expensive research facilities or, a more expanded meaning, the increased productivity gained from collaboration. (Price 1986, 129)

In classical studies Robert.K.Merton (1973) related collaboration to the social stratification of the scientific community, including factors such as age, career opportunities, and scientific status.. In the realm of sociology of scientific knowledge (SSK), collaboration is considered an aspect of the local construction of knowledge, in which research collectives construct local 'truths' about nature and society. Such a process is considered endogenous to knowledge production and driven by knowledge interests. Hence, the objects of sociological study are the roles of different groups within scientific collaboration (and the consequences of their interaction for knowledge production). These two approaches provide important insights into the nature of collaboration and the relationship between collaboration and social factors such as seniority and the production of meaning in collaborative settings.

Katz and Martin (1997) provide six motivational factors that increased research collaboration. These are:

- Escalating costs of conducting fundamental science
- The decrease in the cost of travel and communication, which leads to increased mobility between scientists.

- As a social institution, science depends on interaction and networks to grow. The increased need for specialization in certain fields such as high-energy physics.
- The growing importance of interdisciplinary fields such as biotechnology
- Political factors, such as the growing integration of science in western Europe, that promote cross-national collaboration.

By and large, the previous studies seem to reveal that three sets of factors, economic, cognitive, and social explain collaboration.

1.7.2 Different Loci of Knowledge Production and Collaborative Research

Collaborative research is not a new phenomenon (Godin, 1998). There seems to be a shift in the current meaning and practices of collaborative research. In the earlier decades the academy had its relationship with the 'real world' the on its own terms and choice. As indicated by other writers, (eg, Gibbons et.al, 1994, Stronach and Maclure, 1997: Usher 2000), the political, social and economic conditions in which universities now do research is very different because:

- Researchers and educators, now work in a knowledge sector or industry, which is often described as part of a knowledge economy. In this economy, knowledge is produced and marketed in a many different locations with growing number of participants. Universities therefore have become only one of a number of 'knowledge' producing organizations.
- As researchers and educators, we experience not only the devaluing of disciplinary knowledge and a permeability of disciplinary boundaries but also the breakdown of institutional boundaries.
- 'Real' world imperatives drive knowledge production. The rewarding of research that focuses on problem- problem-solving projects is characteristic of the increased demand for applied knowledge.
- An increase in collaborative research partnerships in which industry contributes to the research costs, and increasing pressure on universities to view their work in commercial terms.
- Increasingly academics, like other employees, align their individual learning goals with those of the organization and increasingly academic work is managerial.

1.7.3 Research Collaboration and Networking

Networking is the new of form of academic research from the point of view of scientific collaboration. This is propelled by the fact that there is a shift in research from an individualistic, disciplinary-based and place-bound ideal towards a collective, problem-oriented and multi-organizational activity. This, according to Mats Benner (2000), is based on an assumption that knowledge is produced and utilized simultaneously within the academic and practical fields and with multiple interaction in the knowledge-producing process as mentioned in the "Triple Helix" or "Mode2" by Leydesdorff (2000) and Gibbons et al. (1994).

Collaboration- Push Perspective emphasizes in the fact that collaboration and networking in general is driven by new models of funding and the collaboration pull perspectives says that collaboration is a universal phenomenon of academic research, explained by the quest for complementarily competencies in the research process, the need for sharing equipment.etc.

Contemporary scholars have distinguished trust, stress, documentary practice, conflict, and perceived success as critical dimensions in knowledge production.

According to (Zabusky 1995) scientists engaged in interorganizational research collaborations are often exposed to high levels of stress because of:

- Complex technology demands,
- Unclear or changing social arrangements
- Need to coordinate geographically dispersed groups, and
- Clash of interests
- Ambiguity in the distribution of authority and
- The pressure to perform according to expectations of funding agencies as well as time constraints.

The last factor is important from the point of view of time as a critical resource in working together. Pressure from funding agencies and participating institutions to perform within tight budgets and under time constraints impart major degree of stress.

Communication demands are intense among large projects involving different organizations requiring both informal and formal communication. According to Latour (1987), the resources involved in projects, documentation in form of notes, memoranda, proposals, plans necessitate a must transfer of data in a faster mode prior to the appearance of published results. Such a documentation practice forms the backbone of the work organization of science.

1.7.4 Structural Dimensions of Collaboration

Orbuch (1997) recognized the role of accounts as explanations of social behavior and social events in sociology. The reconstruction of such accounts depends heavily on the preservation of written documents. Documentary practice as the generation and preservation of inscriptions is essential for the work organization of scientific research and in the form of artifacts; such practice constitutes the "social memory" of collaborations in science.

The author identified seven major structural dimensions of collaborations: Project formation, magnitude, interdependence, communication, bureaucracy, participation, and technological practice. Briefly describes:

Project formation: Collaborations have a variety of origins. In some, one sector is dominant, both in origin and constitution. Others encompass academic, governmental, and private sectors. The role of preexisting relationship varies as well as supervision and funding agency involvement.

Magnitude: Some collaborations are larger than others, in terms of the number of organizations, individual participants, subcontracts, graduate students, teams and costs for personnel.

Interdependence: Data sharing, the autonomy of organizational teams with respect to instrumentation, and the analysis of joint data distinguish collaborations in terms of the interdependence of their constituent social formations.

Communication: Communicating the actors in the collaborative association, and with that of the sponsor of the collaborative project.

Bureaucracy: The degree of bureaucracy is a fundamental aspect of organizational structure.

Participation: The rank of the participants in the colloaboration ranging from the graduate students to that of the head of the respective scientific institutions.

Technological Practice: Collaborations always use technological devices but vary in the extent of their dependence and the forms of use. Characteristics of these uses allow to distinguish a broad array of factors that may be designated the technological practice of the collaboration. Some design and own their own equipment. Some of these instruments are state-of-the-art innovations. Instruments may be changed during the course of the project. Technological practice is not merely instrumentation- it also includes the management of topics and checking the results.

The present study focuses exclusively upon factors like easier and less expensive communication (e.g., fax and electronic mail) that seemed to have increased rates and extent of collaboration which in turn has consequences for the productivity of scientific groups. Productivity in ideality is the sole condition to shape the reward structure of science

1.7.5 Social Significance of Collaboration

One of the most significant developments in twentieth century science is the proliferation of projects that require the resources and expertise of multiple teams of researchers. Indeed, the growth of big science is not simply an increase in the scale of research, measured by exponential growth of the number of scientists and the number of their publications (Desolla price 1963), but an increase in collaborations involving a variety of different institutions. Such interorganizational efforts are found in all areas science that requires significant resources and complex instrumentation.

Sociologists, Historians and Anthropologists demonstrated that the examination of collaboration helps understand processes involving external relationships, consensus formation, and cultural construction. Owing to the financial requirements, risk, and visibility of many research projects, decisions about resources and participation cannot be left to a single scientist or even a science manager. In some cases collaboration encompasses a lager proportion of researchers in a given specialty, their establishment can have a major impact on the consensus formation of genealogical change in the field (Knorr-Cetina 1995). Since instrumentation brings together large number of participants-physically or electronically – subcultures may develop with implications for stratification processes. (Traweek 1988)

Collaboration is important sociologically, because modern production and services have increasingly become knowledge based, with the new knowledge often being created as a result of collaborative endeavors. Efforts to empirically describe and theoretically explain this recent form of social organization have just begun in social studies of science (Zabusky1995, Schild 1997; Knorr-Cetina 1998).

Manichean (1993) with respect to biosciences, proposed 3-fold classification based on reason based on reasons for collaboration:

- To promote an efficient division of labor
- To enhance credibility and
- To build community

In this work we examine the nature and pattern of collaboration among the organic chemists in India and more importantly the relationship between collaboration and productivity.

1.8 Work group size and Productivity

Scientific research is a collective activity with a division of labor among the members of the research group. The advent of project mode of production of knowledge necessitated the participation of groups of students working with a scientist in the production of knowledge. These graduate and postdoctoral students are productive and they acquire the values and norms that will allow them to participate fully and interactively as scientists. With effective socialization they learn to do 'good science'. Thus there have been a number of studies that indicate the importance of work-group characteristics for the development and productivity of scientists in both academic and industrial settings(Louis et.al.2007), mix of higher professors and full time researchers leads to higher rates of productivity (Carayol&Mat,2004) and hiring new and productive staff increases productive members (Lacetera,Cockburn,Henderson 2004). In addition, other studies also suggest that research productivity increases with group size(Bland&Ruffin,1992;Dundar and Lewis,1998; Johnston 1994) and the relationship is moderately curvilinear (Jordan, Meador & Walters,1988). In our study we considered the average size of the research group over a period of time comprising of doctoral and postdoctoral students to examine the impact on the productivity of the scientists.

1.9 Framework of the Study

1.9.1 Organizational Contexuality

In Critical Social Theory, "social" in the term "social action" refers to the orientation of a person's action to other individuals and to the action being embedded in an organizational context. Through its social and institutional structures, the organizational context defined, for all organizational actors, the possibilities and potentialities and potential for social action. In everyday interactions, an organization's policies, norms, and resources, serve to enable, constrain, and sometimes outright determine what is proper and improper, and to lend meaning to the actions of individuals.

The organizational context also defines the power, authority, and status relationships of the individuals within it. However, as intelligent and knowledgeable agents, organizational actors can, within, limits, choose to act in accordance with or against organizational norms. To well- socialized actors, the organizational context is taken for granted store knowledge or a set of pre-interpreted patterns of meaning about the organization.

The organizational context serves as a reference schema that enables actors to act and to interpret the actions of others. As actors mediate action situations, they draw upon these stocks of knowledge, as well as material nonmaterial resources of the organization. In executing social actions, actor replies upon the fact that he or she shares every aspect of the organizational context with the other actors involved in the action situation. Thus contextually of social action has numerous practical consequences for daily organizational life and for researchers who observe it. (Ngwenyama 1991)

Communication

Communication is the exchange of information between individuals by means of a common signal system. Scientific communication can be defined as the combined processes of presentation, delivery and scientific information in human society.

Leah A.Lievrouw and Kathleen Carley (1990) defined:

> ***Communication process***: as any activity or behavior that facilitates the construction and sharing of meaning among individuals, that they consider to be the most useful or appropriate in a given situation. For example, the presentations of research findings in a journal as a publication.
>
> ***Communication Structure***: as a set of relationships among individuals who are linked by the meanings they construct and share. For ex, the research front, invisible college.

Computer Mediated Communication

A formal and informal communication system mediated by the use of computers and e-mail technology

Collaboration

Collaboration may be defined as "production interaction between two or more individuals with a conscious goal of producing a concrete output. (Haribabu 2002)

Productivity

A numerically measurable output by the scientists during a particular time period.

1.10 Statement of the Problem

Scientists are papyrocentric in the sense that they want to publish their results as ***quickly*** as possible. In the organization of science, publication is linked to recognition and rewards. Thus, the accruing of rewards is linked to productivity of the scientists and productivity is taken, as the basis for reward system. Differential allocation of rewards results in stratification.

Science in India is carried out in various settings: Universities, institutions and national laboratories that differ from each other in terms of organization and funding. These factors stratify them in terms of access to resources for research including technology for communication. The "**haves**" of resources tend to have higher rates of productivity and visibility compared to the **"have-nots**". Differential access to and control over resources seem to account for productivity differentials among scientists.

The study concentrates on the relative contribution of age, area of specialization, training, and access to communication technology, collaborative research, and the organizational context as the conditions for productivity of scientists.

Examination of various organizational aspects of science in India like that of the context of production of knowledge, communication, collaboration pattern, productivity, and reward system across different organizational settings will throw light on the social dynamics of science in India A study of the new technologies such as Information Technology, and computer mediated communication technologies will be useful in understanding national and international productive linkages.

The study focuses on the communication in science as an organized system with formal and informal structures like the visible and invisible colleges. The study aims at understanding the structures of communication among scientists. . It also aims at understanding the nature and extent of collaboration among scientists and their scientific groups. The project addresses the questions like what factors are responsible for making collaborative efforts successful.

The study considered productivity as the consequence of the above and the significator for recognition and reward system in Indian science.

1.11 Research Questions

1. How and in what ways the stage of cognitive development of a field, organizational context of postdoctoral training and research, and individual characteristics influence productivity?
2. What role does the context of the organization play in contributing to the productivity of organic chemists?
3. To what extent does access to and control over technology, communication, especially electronic communication instruments influence on collaboration and consequent productivity of scientists?
4. To what extent electronic communication removes/perpetuates inequalities in science at the level of individuals in scientific community and scientific organizations.
5. What could be the implications for a national science policy in the context of changing context of production of scientific knowledge?

1.12 Objectives of the Study

1. To describe the influence of the structural factors like age, duration of organizational affiliation, academic training, area of specialization, on the productivity of organic chemistry community in India.
2. To understand the processes like communication and collaboration among organic chemists in India.
3. To examine how and to what extent the above mentioned factors influence productivity of organic chemists in India.

1.13 Hypothesis

The study hypothesizes that academic training, differential access to and control over technological infrastructure, institutional affiliation are directly related to differentials in scientific productivity.

In the present chapter, an attempt has been made to explain the research problem under study in the light of the review of existing literature and concepts to be operationalized.

In the second chapter, we shall discuss the methodology adapted for collecting data, size of the sample, selection of the respondents, response rates, introduction of the variables selected for different levels of data analysis.

In the third chapter, we shall examine evolution of chemistry as a discipline, focusing on the professionalization of chemistry. We shall trace the historical development of Organic chemistry as a research area explained by various theories in chronological manner. We shall also discuss the nature, salient features, and families of compounds of organic chemistry and the contemporary taxonomy of the sub-discipline of organic chemistry.

In Chapter 4, we shall describe the profile of the respondents both at aggregate and at individual levels. The profile includes, age, education and training, current organizational affiliation, professional experience before joining the current organization, resources for research and mobilization strategies, research specialization, technological resources and Productivity levels.

In Chapter 5, we shall attempt to explain the productivity differentials among organic chemists in the Indian context. We shall examine the relative contribution of factors we considered in our study to account for the productivity levels among organic chemists.

In chapter 6, section- I, we shall examine the nature and pattern of communication in Indian science. We shall examine the shift to electronic mode of communication, and the needs the electronic mode of communication serves the scientists in the production of knowledge. In the second section of the chapter we shall examine the collaborative pattern among patterns according to age, organizational affiliation and to what extent collaboration contributes to the productivity levels of the scientists.

In chapter 7, we shall provide a statistical explanation for the variability in productivity levels of organic chemists through correlation and regression analysis.

In chapter 8, we shall summarize the findings of the study to conclude with suggestions for implications to scientific community in India.

CHAPTER TWO

METHODOLOGY

Introduction

This chapter provides with the rationale for selecting organic chemistry as the discipline for study and the organizations where the study was conducted. It describes the methodological approach employed with reasons and the tools and techniques of collecting the data in the study.

2.1 Selection of the Discipline for Study

Training in Masters Degree in sociology in the area of research methods added a systematic outlook to my natural inquisitiveness to 'observe' things as social facts and external reality. My inquisitiveness was nourished, as I was a non-participant observer among my friends who were doctoral students in the chemistry department of the university of Hyderabad. It is indeed, in their laboratories that I came across, aspects like publication, postdoctoral training, work culture in terms of research groups, collaboration and the use of electronic mail mode of communication for the purpose of research like, 'talk' 'chat' 'ftp', sending manuscripts 'online' etc.,. It nurtured my interest in the idea of what are the determinants of productivity in the Indian scientific community. Hence I chose to work on the Organic Chemists Community in India.

Besides, my supervisor's work on Production of knowledge, interaction and collaboration among molecular biologists was a major influence to decide upon to work on the community of organic chemists.

My enrolment into doctoral program was parallel to the advent of information technology in the Indian context. Communicating with people via-email in an instant mode was a new experience. Most of my friends were doctoral students in the Department of Chemistry where I used to spend nights in the laboratories to 'download' literature related to research topic.

In the public sphere, the discourses about the new modes of communication drew my attention in observing the phenomenon and especially the communication through 'space of flows' (Castelles 1989), the new way of networking and communication as practices and social processes. I started to raise questions on how scientific institutions were orienting themselves to new interacting modes, and how these modes of communication diffused within the Indian institutional context, where state has become the major promoter of this information mode of development. Thus, this book also explores this new space of communication and consumption of these technologies and their influence on the productivity of science personnel, among different organizations and institutions.

2.2 Selection of the Organizations

Scientific research in India is performed in different organizational settings, majority of them being under the state and a few private R&D institutions. We identified the following category of institutions.

- Universities of two kinds. University established by the Central Government and Universities established and maintained by the state or regional governments.
- Indian Institutes of Technology,
- National Laboratories established and maintained by the Council for Scientific and Industrial Research and
- A completely autonomous Deemed University.

2.3 Selection of Respondents

We identified respondents from the Website of Indian Association for Organic Chemists and also scientists recruited in the Departments of Organic Chemistry in the selected organizations. In all 91 scientists were contacted. We collected data from the scientists whoever have agreed to participate in the study by adopting to snowball method of sampling. Out of 91, 53 agreed to participate in the study. (See Table.2.1)
The data were collected for a period between 1996-2000 for about two years.

Table.2.1: Sample size and response rate

ORGANIZATION	NO. OF SCIENTISTS	NO.WHO RESPONDED
STATE UNIVERSITIES		
OSMANIA UNIVERSITY	8	5
ANNA UNIVERSITY	5	3
MADRAS UNIVERSITY	5	3
TOTAL	**18**	**11**
CENTRAL UNIVERSITY	**10**	**10**
IITs		
CHENNAI	9	6
KANPUR	9	4
MUMBAI	10	3
TOTAL	28	13
NATIONAL LABS		
IICT HYDERABAD	28	12
CLRI CHENNAI	4	2
TOTAL	**32**	14
DEEMED UNIVERSITY	**9**	**5**
TOTAL	**91**	**53**
RESPONSE RATES	100%	**58.24%**

2.4 Instruments of Data Collection

The data were collected from the scientists by using both quantitative and qualitative instruments. I used questionnaire for collecting quantitative data. Questionnaires were sent through surface-mail in the first round of data collection. Only one scientist from a state university responded. Hence, I visited the laboratories of the scientists selected for the study and collected quantitative and qualitative data almost simultaneously. Three scientists from the deemed university insisted that I should send the questionnaire by e-mail as they were busy at that time and after doing so only one responded. I asked the scientists to provide me with their bio data and complete list of publications. While some scientists agreed and handed over, some refused saying that their bio-data is not a document for public display.

The qualitative instruments of collecting data helped me to get more into the details of the responses. I relied upon unstructured in-depth interviews with scientists who allocated time for me and discussions with checklist with those who could not spend much time for an extended in-depth interview. The duration of the interviews ranged between half-an-hour to two and half-hours.

2.5 Plan of Analysis

In Sociological research, we came across works, which used either quantitative or qualitative method to study a phenomenon. By being exclusive these methods do not seem to provide comprehensive understanding of the phenomena. In the present work, qualitative and qualitative methods are viewed as complementary rather than rival camps. Zelditch (1962, p.576). Thus, I adopted a convergent methodology, the triangulation approach which,

- Helps organizational researchers to improve the accuracy of their judgments.
- By employing multiple methods, allows researchers to improve the accuracy of their judgments.
- Is largely a vehicle of cross validation when two or more distinct methods are found to be congruent and yield comparable data.
- Advocates multiple methods to examine the same phenomenon
- Is effective on the premise that the weakness in each single method will be compensated by the counter-balancing strengths of another.

2.5.1 Quantitative Analysis

I analyzed the quantitative data at three levels. At the first level, I tabulated data by constructing univariate and bivariate frequency tables. At the second level, I employed partial correlation analysis of the independent variables with the dependent variable and at the level three of multivariate analysis, I developed a multiple linear regression analysis model on the basis of the conceptual framework. The variables considered in the analysis of the data are:

2.6 Variables

Dependent variable: Productivity (PROD)

Though the scientists expressed limitations of a definition that restricts productivity to the number of research publications, there seems to be an agreement that in Indian context, there is emphasis on the quantity. For this reason, for quantitative analysis we considered the number of publications as the indicator of productivity. As it was also difficult to collect complete list of publications by scientists, I considered the total number of publications between the years 1996-2000 as the productivity of scientists.

Independent variables

1. AGE-
   AGE
2. POST DOCTORAL TRAINING DURATION ABROAD
   PDFADU
3. POST DOCTORAL TRAINING DURATION INDIA
   PDFIDU
4. DURATION OF ASSOCIATION
   DURATION
5. VOLUME OF COMMUNICATION
   VOLCOM
6. NUMBER OF COLLABORATIONS
   NOCOLLAB
7. AVERAGE SIZE OF THE RESEARCH GROUP
   GPSIZE

The independent variables 2, 3, 6 and 7 are self-explanatory and hence, we define the independent variables age, duration and volume of communication.

- AGE- Biological age of the scientists in terms of years.
- DURATION-The number of years a scientist is associated with the current organization in which he is employed.
- VOLUME OF COMMUNICATION- Total number of e-mail messages sent and received by a scientist in a week related to research.

Science is a paradigm-directed social action and scientists are actors who share the beliefs and cognitive orientations associated with the paradigm and social/moral norms. In this context, it is important to understand scientific activity and notions of productivity from the point of view of the actors i.e., scientists. For this purpose, in addition to employing statistical analysis of variation in productivity levels and accounting for the variation in terms of a set of independent variables, in this study, I employed phenomenological perspective to understand the productivity from the point of view of scientists through in-depth interviews. Most of the in-depth interviews were held at the laboratories of the scientists. The visits also enabled me to observe the laboratories- in terms of equipment, spatial aspects such as connectivity to library. Thus as a part of this study detailed in-depth interviews were conducted with the scientists.

Qualitative Analysis of interviews and discussions has been used to supplement and enrich the interpretation of the quantitative aspects and also to understand the dynamics of collaboration and communication.

CHAPTER THREE

ORGANIC CHEMISTRY: AN OVERVIEW

Introduction

A sociological study of scientific community in a given discipline should attempt to understand the concerns of the disciplines, paradigm shifts in the concerns over time. A sociologist should be familiar with the cognitive concerns of a community.

In other words, the object of the study, sensitizes sociologists (Knorr-Cetina, 1982) and he/she is expected to understand the dynamics of the cognitive development in a given specialty. As this study is regarding the community of organic chemists it is pertinent to provide an overview of the cognitive concerns of organic chemists over time. Moreover, understanding these shifts is useful to identify research specialties, their growth and productivity of scientists across different age groups in the study.

The specialties in organic chemistry have different roots of evolution, branching and expansion. With the ever-expanding nature of disciplines along the lines of the demands of the goals, set either by community or by the wider society including market.

Often the history of chemistry has been written on national lines. The present study traces the development of chemistry, especially organic chemistry and the evolution of its sub-disciplines in across disciplinary and geographical boundaries in a historical global context looking for roots of communication media among the members of the scientific community.

Social scientists and scientists who wrote extensively on individual disciplines, more specifically on chemistry have adopted different approaches such as:

- *Internalist history of chemistry*: science done without considering the historical context of that science.

- *Externalist history of chemistry*: appraising the importance of socio-cultural context in which the discipline arose and branches and so on.
- *Biographical approach*: discourse analysis of the biographies of scientists, as a biography as a narrow subject focuses on a single individual, but in other senses, it can provide an opportunity for a much broader treatment of the history of chemistry.

The present study attempts to elicit information required for a historical overview of organic chemistry adopted on the basis of all the three approaches mentioned above, to provide a holistic picture of the subject understudy.

3.1. Chemistry in General

Historically human beings have been trying to understand the very basis of life—starting from speculation in vacuum to that of genetic mapping of the day… In the process, all along the line of evolution of the knowledge and technological systems there developed two discourses of pertinence namely, biology -the scientific study of living things. To understand the nature of biological processes it is essential to understand the reactions of carbon compounds.

Chemistry has been called the science of what things are. Its intent is the exploration of the nature of the materials that fabricate our physical environment, why they hold the different properties that depict them, how their atomic structure may be fathomed, and how they may be manipulated and changed.

3.1.1 Professionalization of Chemistry

One of the achievements of Lavoissier during the 18th century was to put chemistry on the map of knowledge as a model experimental science. Hence, in the nineteenth century many more people wanted to study chemistry. Some learned, the subject from books, others attended lecture courses offered by practicing chemists as part of scientific movement in Western Europe (Ben-David 1972). There had been a surge of interest in the popularization of science, which must have included chemistry. For example, the young Michael Faraday taught himself the rudiments of the subject from Mrs Marcet's Conversations on chemistry before receiving a ticket to attend the famous lectures of Humphry Davy at the newly established Royal Institute in London.

Meanwhile, in France, chemistry was one of the subjects taught at the Ecole Polytechnique, founded in 1794. There were ambitious plans to construct several chemical laboratories where the students could carry out experiments. Certainly this was one of the first institutions where practical chemistry was taught. Among the early student was Gay-Lussac, who then became Berthollet's assistant before himself being appointed to the chair of chemistry at the chemistry at the Ecole Polytechnque. Thus the career of Gay-Lussac provides a study for the early professionalisation of science. Chemistry lent itself particularly well to practical applications in industry and in analysis, so that there were many employment opportunities for chemists than for physicists.

In England the professionalisation of chemistry is largely synonymous with the foundation of the Royal Institute of Chemistry in 1887. An important preliminary problem to sort out was to identify real chemists who had received training. In this the foundation of the Royal College of chemistry in 1845 marked a crucial step. The work of the college owed much to its first Professor of chemistry, A.W.Hoffman, a former student of Lie big. One of the major research tasks for the future will be a systematic study of Liebig and his students. Liebig himself had helped to provide a model for laboratory instruction in chemistry in his laboratory at Gissen. His students were of enormous influence and there are many questions to be answered on the basis of future research in this field.

Linus Pauling, a pioneering father in his famous "The study of chemistry'' (1960) propounds that,Chemistry has two main aspects: Descriptive chemistry-the discovery and tabulation of chemical facts, and theoretical chemistry, the formulation of theories that, upon verification, unify these facts and combine them into a system.

In all these, there have been identified three major branches namely, Inorganic, Organic and Physical Chemistry. The present study is a case of Indian organic chemists.

3.2 Organic Chemistry

Many organic compounds were known to and used by primitive/ancient people. Humans have exploited the use of many organic compounds mainly extracted from plants over the past …years for fuel, dyes, medicine and drugs, foods, especially development of

1. Processed food in order to extend the nutritional value of the through winter, traveling and hard times and

2. Pharmaceuticals through the use and study of herbal medicine. Many drugs used today are based directly or indirectly on such traditional knowledge.

The salient feature of organic compounds is that they contain carbon and the best present definition of organic chemistry is the chemistry of carbon-containing compounds. Yet the definition of organic chemistry has been changing over a period of time owing to its ever-expanding scope.

A striking feature of organic chemistry is the vast number of organic compounds known and the fact that potential number is virtually limitless. Over two million different compounds have now been characterized and every year tens of thousands of new substances are added to the list, either by discovery in nature or by preparation in the laboratory.

3.3 The History of Organic Chemistry

Although organic reactions have been conducted by man since the discovery of fire, the science of organic chemistry did not develop until the turn of the eighteenth century, mainly in France at first, then in Germany, later on in England. By far the largest varieties of materials that bombard us are made up of organic elements.

The beginning of the Ninetieth century was also the dawn of chemistry, all organic substances were understood as all being materials produced by living organisms: wood, bone, cloth, food, medicines, and the complex substances that configure the human body. Inorganic material was believed to come from the Earth: salt, metals, and rock, just to name a few. Because of the human's wonder of natural life, organic materials were believed to possess an enigmatic "Vital Force." Thus organic chemistry was separated from inorganic chemistry, and it became separate field of science. By the turn of the nineteenth century, the "Vital Force" theory was immensely discredited, but this branch of science still stayed separated from inorganic chemistry.

Other famous experiments also proved that the vitalism theory was wrong. In 1845 Kolbe synthesized acetic acid, the chief component in vinegar, in a flow of reactions starting with Carbon, the experiment is demonstrated better defined since acetic acid (C6H4O2) is a carbon-carbon bond. The theory of vitalism, like many other scientific theories, disappeared slowly under the weight of accumulated evidence rather than as a consequence of any one brilliant and enlightening experiment.

3.3.1 Structural Theory

The 1860s mark the beginning of the second major period of growth in the organic chemistry field. The development of a detailed picture, by using pure reasoning of both atomic organization and the shapes of molecules stands as a great milestone of the development of human intellect. At almost the same point in time, Kekule in Germany, and Couper of Scotland suggested that atoms in molecules are fused together by bonds. Their theory was that every atom is distinguished by having the same number of bond availability or valence number, wherever that particular atom appears in any compound. The main characteristic of organic compounds is having strong carbon-to-carbon bonds. This was recognized in the theory, and was used to help understand large molecules, possessing many bonded carbon atoms. Carbon is the cement that holds their molecules together. So far, this theory has gone through rigorous testing, and has not been proven inadequate to this day, as of now it is a law. Kekule's and Couper's theoritical formulations were not all without fault; it is suprising that they did not recognize atoms as three-dimensional objects if they were to be understood as true particles of matter in space.

In 1875. Van't Hoff and LeBel proposed their hypothesis of compounds and atoms taking up space. Their hypothesis went as follows: Four bonds of carbon were located at equal angles to each other in space, this would be a ***rectangular tetrahedron***. Immense amounts of proof have been supplied to support this theory, but are not universally accepted. It is believed today that the hypothesis of van't Hoff and LeBel was misdirected. It goes to show science is not always immutable. Theories and laws can be proved wrong. This Hypothesis was no exception; science can adapt to the world around it. After all, the "mission statement" of science is the attempt to understand the world around you, and without change there is no growth. The structural theory is not only a focal point of organic chemistry, but also an amazingly simple idea. It states that by grasping that each carbon atom forms four bonds, tetrahedral arranged in space, we are able to map the architecture of even the most complex molecules. Hence, even though the molecules are too minuscule to be seen in most powerful, cutting edge, electron microscopes. Scientists are still able to possess a clear understanding of how a molecule is constructed. Although the atoms may have physical characteristics different from what scientists expected such as, carbon atom being an elliptical shape, or the bonds may not line up in a compound as neatly as we envisioned them. Nevertheless, the truth of their basic physical architectural hypothesis has been substantiated literally millions of times by successful outcome of prediction. The statement demonstrates the power of the theory that there has been no chemical observation that cannot be basically understood by structural theory. Finally, although structural logic is extremely rigorous, it involves no mathematics. Unlike most sciences of equal complexity, much of organic chemistry is conducted without the use of formal math beyond elementary levels.

3.3.2 The Electron Theory

The third phase in the history of organic chemistry ends with the description of chemical bonds as electron pairs, Lewis came up with this in 1917. Although a great amount of chemical reactions were already known and in active use to synthesize organic compounds into other compounds, only with this understanding of the nature of a chemical bond did a clear reason of the nature and mechanism of chemical reactions begin to appear. This will be clear when one realizes that the transformation of one molecule to another, a chemical reaction, requires the breaking of some bonds and the making of others. This process could not be understood until one knew what a bond is.

Thus if the nineteenth century was devoted to unraveling the fixed structures of molecules, the twentieth century was devoted to the study of their transformations. The study of science and more specifically the study of organic chemistry became an on going affair. In the scientific community one never rests, there is a continual stream of experimentation and the desire to explore new realms. The cutting edge in science is now grounded in the medical field. Today we manipulate genetic codes, the building blocks of life.

At the root of the sate-of-the-art biochemistry and bio-informatics which aim at improving human life at the molecular and cellular level, Organic Chemistry seems to be an important discipline by virtue of its historic criterion to classify anything that deals with carbon compounds as organic. It is carbon that forms one of the essential elements of life along with Oxygen and Hydrogen.

Thus, it appears that in order to understand the nature of biological processes that constitute our existence, it is essential to understand the reactions of carbon compounds, the essence of Organic Chemistry.

The things we have learned over the years allow us to build those bridges to the future, a future that might see an improvement in the human condition.

3.4 Families of Organic Compounds

Organic Chemistry is the largest branch and fastest growing branch of Chemistry. The vast numbers of organic compounds are classified into "families". Each family consists of compounds that have a chemically active center of the molecule called the family's "functional group". All members of a particular family have similar chemistry because their functional group is the center of Chemical activity. Some of the families include the following:

Alkanes (all single bonds)
Alkyl Halides (involving at least one halogen bonded to a carbon)
Alkenes (at least one Carbon - Carbon double bond
Alkynes (at least one Carbon- Carbon triple bond)
Aromatic Compounds (involving the molecule Benzene, Napthalene, Anthracene, etc)
Alcohols (at least one OH group)
Thiols (similar to alcohols except Sulfur (SH) instead of Oxygen
Ethers (at least one Oxygen single bonded to two carbons)
Thioethers(similar to ethers except a Sulfur atom in place of an Oxygen atom)
Aldehydes (at least one formyl group -CH=O)
Ketones (at least on keto group C=O)
Carboxylic Acids (at least one carboxyl group -COOH)
Amines (at least one Nitrogen bonded to Hydrogen or carbon atoms)
Amino Acids (at least one amino group NH_2 and one carboxyl group - COOH)
Carbohydrates (at least several OH groups and a formyl or keto group)
Organometallics (where we have ionic bonding between a metal and a carbon structure)

There are other families that have been added as the research progressed over the years.

3.5 The Nature of Organic Chemistry

The nature of organic Chemistry has changed greatly since 1828. As mentioned earlier, the scientific philosophy known as "Vitalism" maintained that ***Organic Chemistry was the chemistry of living systems.*** It maintained that Organic Compounds could only be produced within living matter while inorganic compounds were synthesized from nonliving matter. Even the word "organic" comes from the same root as the word "organism" or "organ". However people like Professor Wohler beginning in 1828, determined that it was indeed possible to synthesize organic compounds from those compounds that were considered inorganic. One of the first organic compounds synthesized from basically inorganic compounds was the compound Urea, which is a metabolic product of urine. It was synthesized from Ammonium Cyanate considered a compound produced outside of living matter and therefore considered inorganic. Since then many millions of organic compounds have been synthesized "in vitro" in other words outside living tissue. Wholer's work showed application potential of organic chemistry.

As mentioned above, Organic Chemistry has developed into a branch of Chemistry that focuses upon the carbon containing compounds. It has just recently been expanded to include compounds of Silicon since Silicon is similar in behavior to Carbon being in the same group within the Periodic Table. Given that the main material in which microchips of the computer have as their foundation is Silicon, it is interesting that the main element establishing living organisms should be merged with the main element involved in the inanimate machine world.

A major contributor to the growth of the use of organic compounds since the 1850s has been the availability of cheap starting materials (chemical feed stocks) from initially the distillation of wood and coal, the subsequently petroleum and now increasing, natural gas.

3.6 Organic Chemistry-Contemporary Taxonomy

The 20th century has witnessed a radical shift in scientific approach to understand the process of life and query to understand the basis of life. The contemporary researchers cut across disciplinary boundaries and work in unison to understand life processes. In this process, each discipline and its sub-branches reorient themselves in collaboration with others. As a consequence of the collaboration, different new hybrid sub disciplines emerge. Organic chemistry is not only a no exception but also a pro-hybrid, pro-change oriented discipline. For the purpose of the present study, we may categorize the sub-branches of Organic Chemistry as:

1. Structural analysis 2. Organic Synthesis and 3.Reaction dynamics

3.6.1 Structural Analysis

Structural analysis is common to most organic chemical research. In general structure must be determined every time a chemical reaction is carried out and a reaction product isolated. If the product is suspected of being a compound already known, it is then physically compared with the known material to establish its identity. The problem is very simple since the molecules used in the reaction and the nature of the reaction itself are already known, and thus offer a strong presumption of the product identity, by analogy. If the product is a new compound, the structure must be proved independently. But when new compounds from nature, namely natural products are isolated specific problems arise. The unknown compound is submitted either to known chemical reactions in order to draw structural inferences or examined for certain physical properties, which have been correlated with structural features. The latter, or physical, methods have had a phenomenal advance in the past three decades, notably in the development of optical and nuclear magnetic spectroscopy. These methods can yield in minutes, extensive information about the structures of unknown compounds that previously could be obtained only by weeks or even years of chemical study. Total determination of structure is also possible by means of x-ray diffraction.

All chemistry is in the broadest sense, structural chemistry. In this era of genomic and proteomics, the same can be said for biology to a large extent. X-ray analysis coupled with computational approach helps in understanding the molecular recognition process. Today rational drug design involves molecular recognition.

Drug discovery traditionally has relied more on serendipity than on rational design, especially in the early stages of lead molecule identification. At present, the optimization of a lead compound into a pre-clinical and ultimately a clinical drug candidate follows a more traditional medicinal chemistry approach of exploring of structure-activity relationships by methodical synthesis of analogues, guided by the expertise of the medicinal chemist.

However, there has always been interest in pursuing a rational approach to drug discovery, driven by the opportunity to accelerate the optimization stage or to access greater chemical novelty. In the lead optimization phase, the study of quantitative structure-activity relationships (QSAR), led by the disciplines of molecular modeling and physical property estimation, has had proven value in rationalizing the structural and physico-chemical properties leading to activity.

3.6.2 Organic Synthesis

There is excitement, adventure, and challenge, and there can be great art in organic synthesis. These alone should be enough and organic chemistry will be sadder when none of its practitioners are responsive to these stimuli.
—R.B.Woodword, 1956

At the beginning of its history, organic chemistry was perceived as a branch of natural science, dealing with a specific

a. synthesis of hundreds and hundreds of compounds which never before existed on this planet and have no resemblances to natural compounds and
b. as an instrument for the discovery of and study of natural phenomena, but is is also capable of creating a wide variety of unnatural compounds.

Chemical synthesis is uniquely positioned at the heart of chemistry and its impact on our lives and society is all-pervasive. For instance, many of today's medicines are synthetic and many of tomorrow's will be conceived and produced by synthetic chemists. To the field of synthetic chemistry belongs an array of responsibilities which are crucial for the future of mankind, not only with regard to the health, material and economic needs but also for the attainment of an understanding of matter, chemical change and life at the highest level of which human mind is capable.

The post World War II period encompassed remarkable achievement in chemical synthesis. In the first two decades, of this period, chemical syntheses were developed which could not have been anticipated in the earlier part of this century. For the very first time, elaborately conceived multi step processes assembled several complex molecules, for example, Vitamin A (O. Isler, 1949).

This striking leap forward, which was recognized by the Award of Nobel Prize in chemistry to R.B.Woodword in 1965 was followed by an equally dramatic scientific advance during the past four decades, in which chemical synthesis has been raised to a qualitatively higher level of sophistication.

Today, in many laboratories around the world, chemists are synthesizing at an astonishing rate, complex carbogenic structures, which could not have been propelled by the availability of more powerful conceptual processes for the planning of chemical synthesis and reagents, and the advent of improved methods for analysis, separation and determination of structure. Many talented investigators all over the world have contributed to the collective undertaking of vast dimensions, even though made independently and their ideas and discoveries interact synergistically to the benefit of all.

3.6.3 Reaction Dynamics

The third broad area of research involves the study of the intimate details of the course of chemical transformation, with the aim predicting quantitatively the rate of any given reaction and the nature of its products. This study of reaction mechanism began in the 1920s following Lewis electronic description of the chemical bond, and was actively pursued in England by Robinson and Ingold, in Germany by Arndt and Meerwin, and in the United States by Hammett, Conant, and Lucas. Extending the methods of physical chemistry these researchers and their contemporary successors have determined quantitative rates of reaction and the effects on these of temperature, concentration, molecular variations, and reaction. The considerable success of these studies has not only increased our knowledge of the intimate details of chemical change and greatly expanded the possibilities of synthesis but has also brought a theoretical unity to the whole field of organic chemistry which has the effect of making its principles easier to teach and to learn.

Further during the middle of the 20th century the various branches of Organic Chemistry have specialized further encompassing and assimilating into other branches of life and physical sciences giving rise to numerous sub-branches that address specific parts and problems of different living organisms.

In the next chapter, we shall examine the profile of the organic chemists in our study and attempt to know for a preliminary understanding of the individual and organizational profile of the respondents in the study.

3.7 Summary

Organic chemistry had its origin along with life sciences in the form of chemistry as a discourse to understand natural phenomena like life and health, sought by mankind. It emerged as a natural science as chemistry was professionalized in Western Europe as an academic movement as a part of popularization of science. Establishment of experimental laboratories in France with an eye on industrial application and Royal Institute in England, as an academic discourse paralleled this.

Chemistry emerged as a discipline with facets, i.e., descriptive and theoretical. Both these facets of chemistry gave rise to three primary branches, namely,

- Inorganic chemistry
- Organic Chemistry
- Physical Chemistry

Organic Chemistry, over a period of time was used to extract food and pharmaceuticals from plants. The word '*Organic*' was synonymous with the presence of carbon in the chemical compounds. As a discipline in its history, organic chemistry was initially explained by Vital Force Theory – which, emphasized upon the existence of an unexplained vital force prevalent among the carbon atoms. This was disproved by the discovery of the Carbon- Carbon bond. Later structural theory that explained the atomic structure of carbon characterized by the discovery of tetrahedron structure of the carbon molecule and the discovery of the Benzene structure provided a new interpretation of Organic Chemistry. In the 20th century electronic theory evolved and explained organic reactions by focusing upon the chemical bonds at the electronic level giving a new dimension to Organic chemistry. Today organic chemistry with its vast applications at the molecular, cellular, and electronic levels has emerged as an essential element of Medicine, Biochemistry, Genetic Engineering, and Bioinformatics.

Organic chemistry diversified into many families, basing upon the functional group in the molecules, today stands widely accepted as a research area with organic structural analysis that aids in molecular and rational drug design, organic synthesis- useful in synthesizing organic compounds from natural products and chemicals as well, reaction dynamics.

In the next chapter, we shall examine the profile of the organic chemists included in the work, both at individual and organizational levels.

CHAPTER FOUR

ORGANIC CHEMISTS: A PROFILE

Introduction

This work helps to understand how and in what ways different organizations in which organic chemists are located influence their productivity. The chapter is divided into two sections. In section I, I describe the profile of the respondents both at aggregate and at individual organizational levels.

The profile includes: age, education and training, current organizational affiliation, professional experience before joining the current organization, resources for research and mobilization strategies, research specialization, technological resources and productivity.

In section II, of this chapter we will describe the productivity levels of the scientists to review their performance in terms of the number of publications they have to their credit in the refereed journals. We categorize the scientists in terms of levels of productivity and analyze productivity differentials in terms of organizational affiliation and research specialty.

I shall examine the technological aspect of scientific research in two ways. First, we provide with, a comparative description of the extent of infrastructure available in terms of adequacy at different organizations. Secondly, I shall focus upon technology as a tool for communication with a description of information network management all over the organization and locate the respective departments and research groups in the network. In relation to the information technology management system I also describe the laboratory setting in each of the organizations under study and observe for adaptive variation in the context of incorporation of information technology and computer mediated communication.

I shall, then describe, the inter organizational cooperative endeavor of the scientists in the form of collaborative research.

4.1 Age Groups among Respondents

I considered age as an important variable in the study as the career graph of a scientist varies with time. Typically a young scientist depends on a mentor, in the initial stage of his/her research. Once she/he begins her/his research career depends on the organization for initial support and later establishes herself/himself in the profession and the organization as an independent researcher. I examine at what age do they become independent, mobilize funds for research, acquire infrastructure and also recruit students into the research group?

Figure 4.1: Age Groups among Respondents

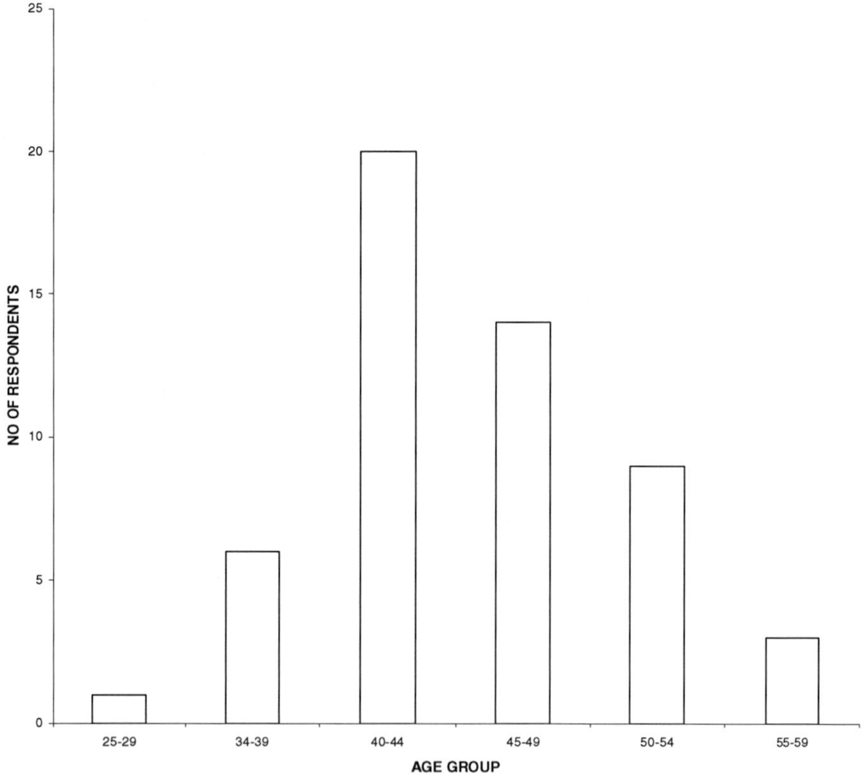

I describe at what age scientists become interdependent in terms of mutual exchange of information, ideas, technology and collaborative research and so on. In the sample, the majority of the scientists belong to the age group of 40-49 years. (See. Fig.4.1).

In the study the age ranged between 28 and 59 years and the average age is 45.07 years. In the sample under study 1.8 percent of the respondents i.e. only one belongs to the age group of 25-29 being the youngest recruited and in the CSIR laboratory. (See Table.4.1)

Table 4.1: Age Groups among the Respondents

AGE GROUPS	FREQUENCY	PERCENTAGE
25-29	1	1.89
34-39	6	11.32
40-44	20	37.74
45-49	14	26.42
50-54	9	16.98
55-59	3	5.66
Total	**53**	**100.00**

In the sample, 11.32 percent of the respondents fall in the category of 34-39. Majority of the respondents falls in the third category, class of 40-44. They constitute 37 percent of the total respondents. In the sample, 26.41 percent of the respondents belong to the fourth category, while 16.98 percent of the respondents are in the class 50-54. Only 5.66 percent of respondents are in the age group 55-59. (See Table.4.1)

4.2 Doctoral Training

The professional life of a scientist starts with his enrolment into the doctoral degree. It is while pursuing the doctoral degree that an individual gets socialized into the scientific community. Under the guidance of a mentor he acquires the necessary conceptual and technical skills. It is in this phase that the student is socialized into cognitive and social norms, values and traditions of research and cognitive and social norms of a scientific community. This phase of socialization into profession occurs both at formal and informal levels. It is at this doctoral level, a student learns in a professional manner about the infrastructure and its uses and is trained to interact with the peers within and outside the research group. The norms of enrolment, training and the level and sophistication of infrastructure may vary from one organization to the other and also influence the performance of the scientist later. We may hypothesize that persons trained by recognized researchers at reputed places will have opportunities to be more productive.

Later, I examine the relative influence, if any, of the doctoral affiliation on recruitment, and performance of the scientists.

Table 4.2: Doctoral Training

PLACE	FREQUENCY	PERCENT
ABROAD	10	18.87
INDIA	43	81.13
Total	53	100.00

I compare the productivity levels of the scientists who pursued doctoral degree abroad, i.e. outside India and in India. While 81.13% of respondents pursued doctoral degree in different organizations within India, a small proportion (18.87%) obtained their doctoral degree from foreign universities, and almost all of them from institutions in the United States of America. (See Table.4.2)

4.3. PhD Training and Current Organizational Affiliation

Further, it is important to see from which institutions respondents got their PhD training within India and abroad and their current organizational affiliation.

It is observed that, 22.64% of the respondents obtained doctoral degree from foreign universities, followed by 20.75% from central universities, 18.86% from deemed university, and from national laboratories and a majority of 28.23% are from state (regional) universities.

Figure 4.3: Current organizational affiliation and training

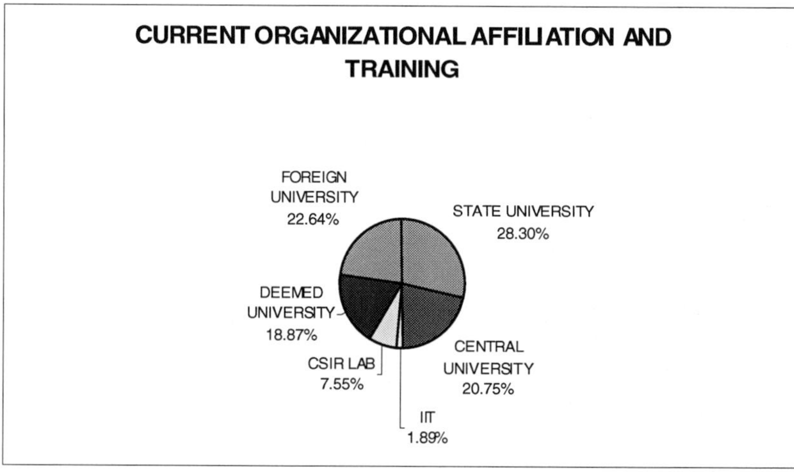

Table .4.3: Institutions of Doctoral Training and Current Organizational Affiliation

ORGANIZATION	DOCTORAL AFFLIATION						
	STATE UNIV	CENTRAL UNIV	IIT	CSIR LAB	DEEMED UNIVERSITY	FOREIGN UNIVERSITY	TOTAL
STATE UNIVERSITY	10	-	-	-	-	1	11
CENTRAL UNIVERSITY	-	2	-	-	4	4	10
IIT	-	6	-	-	3	4	13
CSIR LAB	4	2	1	4	2	1	14
DEEMED UNVIERSITY	1	1	-	-	1	2	5
TOTAL	**15**	**11**	**1**	**4**	**10**	**12**	**53**

Ten out of 11 scientists in the regional universities obtained their doctoral degree from state universities themselves. None from the other categories of organizations have been recruited into state universities. Only one Professor, aged above 50, with having done his doctoral degree abroad, was the head of the department in a state university, at the time of the study (see Table 4.3)

In case of the scientists at the Central University, 40% of them obtained doctoral degree from foreign universities, another 40% from deemed universities and 20% from central universities. None of the scientists who obtained doctoral degree from a state university found placement at the central university.

Among the respondents currently affiliated to the IITs, 46.18 % obtained their doctoral degree from central universities followed by 30.76 % from foreign universities and 23.07 % from a deemed university. Again none of our respondents who obtained Ph.d from a state university found placement at the IITs. The concentration of scientists who obtained their Ph.d from state universities at the same State University indicates that there is high degree of inbreeding at the state universities.

Four scientists in our study working at the national labs, have obtained their doctoral degree from state universities, while two have obtained their degree from central universities and one from IIT and another scientist from a foreign university. In the case of five scientists in the deemed university, three obtained their doctoral degree from state, central and deemed universities respectively and 2 obtained their doctoral degrees from foreign universities.

4.4 Post Doctoral Training

The reputation of the institution where one obtains her/his PhD degree and the academic sanding of the research adviser/supervisor play an important role in shaping one's research career. The institutions are generally stratified according to their standards and resource-endowments both physical and cultural. Obtaining a PhD degree from an institution with high academic/research standards marks out a career-path for a researcher. Further, the academic recognition of the research advisor/supervisor also plays a role in shaping a researcher's career. One's research advisor/supervisor helps the researcher in securing either post-doctoral fellowship at reputed laboratories or an appropriate placement through her/his recommendations, which are sought by the prospective employers. In the following paragraphs we shall look at the postdoctoral training of the scientists in the study.

During the pilot study, I observed that, postdoctoral training seems to play an important role in the life cycle of a scientist. On an average, the respondents spent two years duration of postdoctoral training either in Indian institutions or abroad. Postdoctoral training represents a transitional stage between doctoral degree and the regular appointment of a scientist. I attempted to see why scientists undergo postdoctoral training, to assess the importance of the same in the career cycle of a scientist, especially in terms of employment opportunities and post- employment performance. I also aim at bringing out how different postdoctoral training is from that of doctoral training, and whether post doctoral training is recognized as institutionalized phase of scientist's career and how, different organizations consider post doctoral training, especially at time of recruitment. The study focused on the difference between postdoctoral training outside India and within India. I attempt to bring out whether the two different categories of postdoctoral training have any impact on the performance of the scientist in terms of productivity.

It is observed that a majority (66.03%) of the scientists have undergone postdoctoral training. Postdoctoral training seems to play an insignificant role in state (regional) universities, as only two scientists from state universities have pursued postdoctoral training. The national laboratories also do seem to attach some importance to post doctoral training as 50% from these national labs were postdoctoral fellows before recruitment.

All the scientists except two scientists in the age groups of 40-44, and 50-55 from the central university had postdoctoral training before being recruited. In IITs and Deemed University every scientist has been a postdoctoral fellow before joining these organizations. Thus, it appears that the deemed university, the IITs, and the central university recognize the significance and desirability of postdoctoral training for a good research career. Whereas on the other extreme the state universities do not seem to attach much importance to post doctoral training and the national laboratories stand in the middle of the continuum,

4.4.1 Post Doctoral Training outside India

The organic chemists in the study seem to place premium on post-doctoral training abroad. Thirty-three (62.26%) respondents received training as postdoctoral fellows, research associates in various universities and research institutes. Scientists across different organizations reflect this feature. The duration of postdoctoral training ranged between less than One year to six years. As mentioned above,9 respondents in the state universities and 9 in national laboratories and 2 scientists from the central universities did not have post doctoral fellowships. (See Table.4.4.1)

A majority of the organic chemists in the study in case of scientists 18 (33.96%) postdoctoral training ranged between 2 3 years, followed by 11(20.75%) between 3-6 years. The scientists at IITs have longer duration of postdoctoral training (3.3 years on an average) at the time of recruitment followed by central university (2.62 years), and deemed university with 2.3 years. We have not considered the averages for the state universities and national laboratories as in their case only a small number of scientists received postdoctoral training abroad. (See Table.4.4.1)

Table.4.4.1: Type of organization and duration of postdoctoral training abroad

TYPE OF ORGANIZATION AND DURATION OF POSTDOCTORAL TRAINING ABROAD

PDFADU	STATE UNIV	CENTRAL UNIV	IITS	CSIR LABS	DEEMED UNIVERSITY	Total
0	9	2	-	9	-	20
1	1	-	1	1	1	4
2	1	5	2	-	1	9
2.4	-	-	-	-	1	1
3	-	2	2	3	1	8
3.1	-	-	-	-	1	1
3.5	-	-	1	-	-	1
4	-	-	4	-	-	4
5	-	1	1	1	-	3
6	-	-	2	-	-	2
Total	**11**	**10**	**13**	**14**	**5**	**53**

4.4.2 Post Doctoral Training In India

In this work it is found that a very small number of scientists received post doctoral training in India, i.e., 11(20.75%) (See Table 4.4.2)

In case of scientists from state universities, 4 scientists received postdoctoral training in India. In case of central universities 40% of the respondents were postdoctoral fellows in India for a period of one year each before they moved onto foreign universities and institutes for postdoctoral training. From the deemed university none was postdoctoral fellow in India.

Table.4.4.2: PDF India duration organization wise

	TYPE OF ORGANIZATION					
PDFIDU	STATE UNIV	CENTRAL UNIV	IITS	CSIR LABS	DEEMED UNIVERSITY	Total
0	7	6	12	12	5	42
1	3	4	-	1	-	8
2	-	-	1	-	-	1
3	-	-	-	1	-	1
5	1	-	-	-	-	1
Total	**11**	**10**	**13**	**14**	**5**	**53**

Among the scientists at the IITs, and national laboratories, only a small proportion were postdoctoral fellows in India. It clearly shows that post-doctoral training in reputed foreign universities does influence one's chances of getting employment in a reputed institution in India.

4.5. Funding

The study examines the economic resources available for research and strategies of mobilization by the scientists. We review the extent and status of the recurring and non-recurring in-house budgetary support to the scientists across different organizations and additional sources of such support.

4.5.1 In-House Support for Recurring Expenditure

In publicly funded research institutions like universities and CSIR labs, the budgetary supported is divided into two categories, support for recurring expenditure i.e., for buying consumables such as chemicals, glassware and non-recurring expenditure such as equipment and fixed assets.

Today, science is carried out in an institutional context with substantial financial investment. In the research endeavor, the production of knowledge requires, acquisition of laboratory equipment, chemicals, and infrastructure related to electronic storage, retrieval and dissemination of large-scale information, on a regular basis. Scientists adopt different methods to mobilize financial resources. The organizations in which scientists are employed provide financial support to some extent as in-house support for expenditure, on recurring and non-recurring items. The amount of in-house budget for recurring expenditures varies from one organization to another as they differ in terms of their resource endowment levels. We intended to explore as to what extent the budgetary support allocated to scientists, supports the research activity and how do scientists manage the economic aspect of scientific research.

Figure 4.5.1: % Of in-house recurring budgetary support

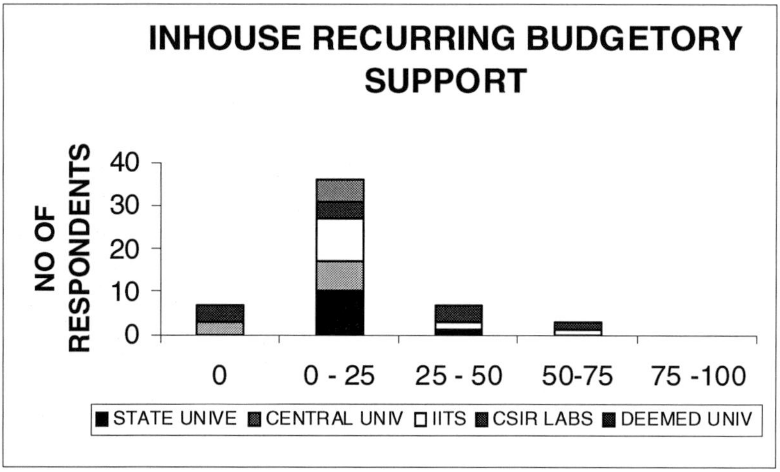

At the aggregate level, 36 respondents in the study mentioned that they receive less than 25% of the required recurring expenditure from in-house sources. Seven scientists, 3 from the central university and 4 from the CSIR lab mentioned that they did not receive any budget support. (See. Table.4.5.1).

Table.4.5.1: % of in-house recurring budgetary support

ORGANISATION	0	0 – 25	25 - 50	50-75	75 –100	TOTAL
STATE UNIVE	-	10	1	-	-	11
CENTRAL UNIV	3	7	-	-	-	10
IITS	-	10	2	1	-	13
CSIR LABS	4	4	4	2	-	14
DEEMED UNIV	-	5	-	-	-	5
TOTAL	**7**	**36**	**7**	**3**	**0**	**53**

4.5.2 In-House Budgetory Support For -Non-Recurring Expenditure

Scientists, in addition to the regular in house budgetary support, also receive one-time grants from the employers to cater to different research needs of the scientists. This again varies from one organization to another.

Table. 4.5.2: % of in-house budget non-recurring

ORGANISATION	0	0 - 25	25 - 50	50-75	75 -100	TOTAL
STATE UNIVE	4	6	1	-	-	11
CENTRAL UNIV	3	7	-	-	-	10
IITS	5	8	-	-	-	13
CSIR LABS	3	8	3	-	-	14
DEEMED UNIV	-	4	1	-	-	5
TOTAL	**15**	**25**	**5**	**0**	**0**	**53**

At an aggregate level, 33 scientists reported that they receive 25% of resources for non-recurring expenditure form in-house resources, followed by 5 scientists with 25-50% support for non-recurring expenditure from in-house resources. As many as 15 scientists reported, that they did not receive any support for non-recurring expenditure from the in house budgetary resources. (See Table.4.5.2)

4.5.3 Status Of in-house Budgetary Support

The scientists in the course of the study were asked to indicate, whether the in-house budgetary support has registered any increase over the past five years, (1996-2000) (See. Fig. 4.5.3). Twenty respondents reported an increase, 11 respondents reported that it decreased and a majority of 22 respondents said that it remained the same. (See Table.4.5.3.)

Figure 4.5.3: Status of in-house budgetary support

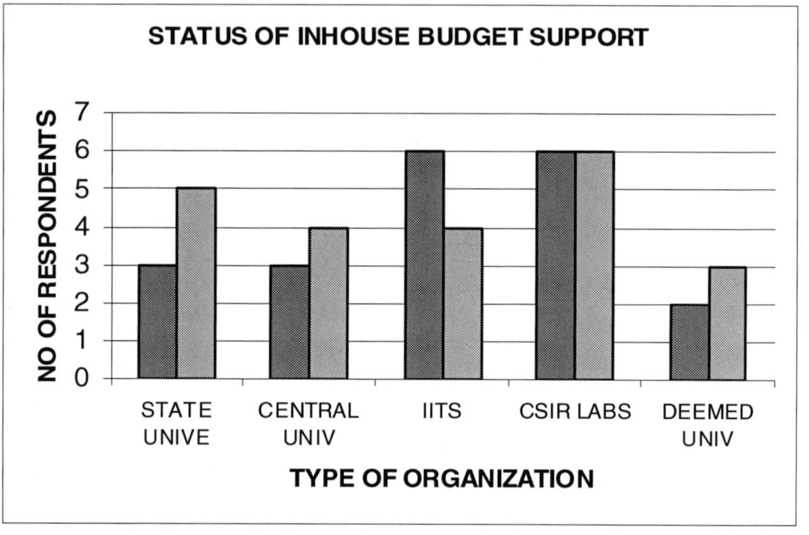

Table. 4.5.3: Status of in-house budgetary support

ORGANISATION	INCREASED	REMAINED THE SAME	DECREASED	TOTAL
STATE UNIVE	3	5	3	11
CENTRAL UNIV	3	4	3	10
IITS	6	4	3	13
CSIR LABS	6	6	2	14
DEEMED UNIV	2	3	-	5
TOTAL	**20**	**22**	**11**	**53**

4.5.4. Alternative Sources: Sponsored and Consultancy Projects

The study reveals that from the organizations, in which scientists are employed, they receive only up to a maximum of 50% in-house budgetary support for research. Scientists seem to mobilize the rest of the economic resources through alternative means like research projects sponsored by pubic and private agencies, and taking up industrial consultancy projects.

It is observed that 71.69% of the respondents undertake sponsored projects from different sources like DST, CSIR, DBT and DAE.

Regarding consultancy projects, less than 50% of the respondents undertake consultancy projects. The deemed university leads in this regard, as 80% of the respondents in the deemed university have consultancy projects, followed by scientists from national laboratories and IITs. In the case of the Central University, less than 40% of the scientists have alternative consultancy projects. (See Table.4.5.4.)

Figure 4.5.4: Sponsored and consultancy projects

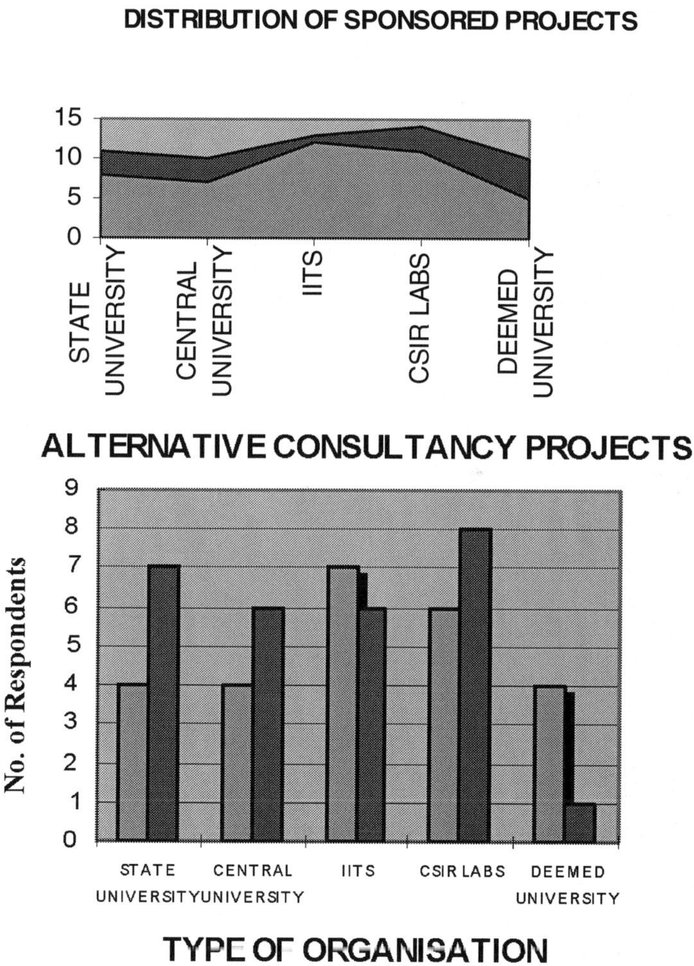

Table.4.5.4: Sponsored and Consultancy projects

ORGANISATION	SPONSORED PROJECTS			CONSULTANCY PROJECTS		
	YES	NO	TOTAL	YES	NO	TOTAL
STATE UNIVERSITY	8	3	11	4	7	11
CENTRAL UNIVERSITY	7	3	10	4	6	10
IITS	12	1	13	7	6	13
CSIR LABS	11	3	14	6	8	14
DEEMED UNIVERSITY	5	5	5	4	1	5
TOTAL	**38**	**15**	**53**	**25**	**28**	**53**

It is clear that there is a variation among the organizations in terms of attracting research funds. The ability of scientists working in various organizations indicates the operation of "Mathew Effect" as described by Robert Merton (1972), according to which disproportionately more resources are allocated to a few institutions where there is a concentration of recognized scientists. In our study we find that, there is a gradation of institutions in terms of allocation of resources. IITs, Central Universities highest positions, where as, the state universities occupy the lowest position.

4.6. Research Area

In chapter 3 we mentioned that in the course of history of organic chemistry new branches and new specialties emerged over time. Exploration into sub-disciplines in a discipline and the convergence of the new sub-disciplines with those of the others is a natural process in scientific research similar to the paradigm shift advocated by Thomas Kuhn (1962). Scientists who switch over to converging areas are perceived as role hybrids (Diana Crane 1969). We examined how and why scientists from different organizations shift their attention to new areas of research and also whether role hybrids tend to be more productive.

From the primary and secondary (curricula-vitae) data it is observed that scientists in our study are engaged in different sub-disciplines of organic chemistry starting from the most traditional heterocyclic compounds of natural products to that of the modern supra-molecular chemistry, an exemplary merger of molecular biology and computer aided rational drug design. We categorized these disciplines into five specialties for the purpose of analysis. (See.Fig. 4.6.1)

Figure 4.6.1: Distribution of scientists in the sub-disciplines

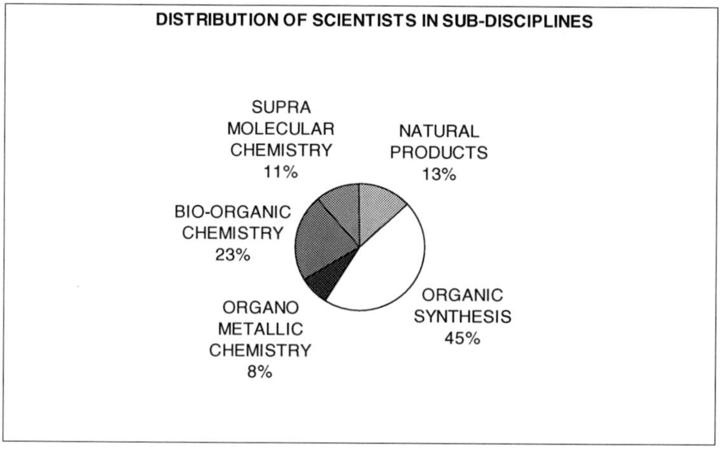

I found that that 46% of the respondents specialize in organic synthesis, 23% in bioorganic chemistry, 7 scientists in natural products, 6 in supra molecular chemistry followed by 4 scientists in Organo-metallic chemistry.

Table.4.6.1: Organic chemists and their specialization

ORGANIZATION	NATURAL PRODUCTS	ORGANIC SYNTHESIS	ORGANO METALLIC CHEMISTRY	BIO-ORGANIC CHEMISTRY	SUPRA MOLECULAR CHEMISTRY	TOTAL
STATE	5	6	-	-	-	11
CENTRAL	-	2	2	3	3	10
IITS	1	4	2	6	-	12
CSIR LABS	1	10	-	3	-	14
DEEMED UNIV	-	2	-	-	3	5
TOTAL	**7**	**24**	**4**	**12**	**6**	**53**

If we look at the specialties of scientists in terms of their organizational affiliation, we notice that, in case of state universities, scientists seemed to specialize in more traditional areas of research, that is, natural products and organic synthesis. From the central university 2 scientists aged above 50 continue to do research in the area organic synthesis, 2 in Organo-metallic chemistry, and 3 each in bioorganic and supra molecular chemistry.

In the IITs category, around 50% of the scientists are engaged in bio-organic chemistry, followed by 4 scientists in organic synthesis, 2 in Organo-metallic chemistry, and only one in natural products. In the case of CSIR national laboratories, 10 out of 14 respondents specialize in organic synthesis, 3 in bioorganic and only one in natural products. The deemed university depicts a complete polarization, as only 40% of the scientists pursue research in the traditional organic synthesis and 60% in the latest modern supra- molecular chemistry.(See.4.7.1). It shows that only scientists affiliated to reputed and well-endowed institutions are engaged in research in the frontier areas of organic chemistry.(See Table.4.6.1)

4.6.1 Age and Specialization in the Area of Research

The study aimed at finding out variation in research interests, in terms of age of respondents, and the reasons for such variation.

I observed that, 64.15% of scientists in the age group 40-49 and they comprise the majority of each sub-discipline. Four out of 7 scientists engaged in natural products fall in the category of 40-49 with the other 3 being above 50 years of age. None in the younger age group, i.e. below 40 years are pursuing research in the area of natural products.(See Table.4.7.2). It indicates that newer areas attract more young scientists, when compared to older scientists. Thomas Kuhn (1962) points out that it is younger scientists and those who are new to the field tend to make more significant contributions in new fields of research. (See Table 4.6.2)

Table.4.6.2: Age and research specialization

AGE-GROUP	NATURAL PRODUCTS	ORGANIC SYNTHESIS	ORGANO METALLIC CHEMISTRY	BIO-ORGANIC CHEMISTRY	SUPRA MOLECULAR CHEMISTRY	TOTAL
25-29	-	1	-	-	-	1
35-39	-	3	-	3	-	6
40-44	3	4	1	8	4	20
45-49	1	8	3	1	1	14
50-54	2	6	-	-	1	9
55-59	1	2	-	-	-	3
TOTAL	**7**	**24**	**4**	**12**	**6**	**53**

Thus, as a research specialty, natural products seem to be a domain of relatively older scientists. Except for one, all other 12 scientists aged above 50 are practitioners of research in natural products or synthetic organic chemistry. Interestingly 50% of the synthetic organic chemists, all Organo-metallic chemists, 75% of bio-organic chemists and 83.6% supra molecular chemists fall in the 40-49 years age group. The data indicates that for older scientists it is difficult to switch over to frontier areas as it involves investment of time, energy and resources to mover over to newer fields and start making contributions.

4.6.3 Extent of Shift of Research Area

Science is an ever-expanding horizon. Different specialties emerge in a given discipline or as a result of cross-fertilization of ideas between. Some of the new branches seem to offer more opportunities for original research and for maintaining higher level of productivity and consequent recognition. Scientists tend to pursue doctoral program in one area of research and later may shift to another for the reasons mentioned above.

In this study, I aimed at, finding out whether scientists change the area of research after their doctoral research. I also tried to explore whether those who continue in the same area of research tend to be more productive. I asked the scientists to indicate whether there has been a shift, and if there was a shift, the extent of the shift. The extent of shift ranges from very small extent to a very large extent through small extent and the large extent. (See Fig. 4.6.3)

Fig. 4.6.3: Extent of shift of research area

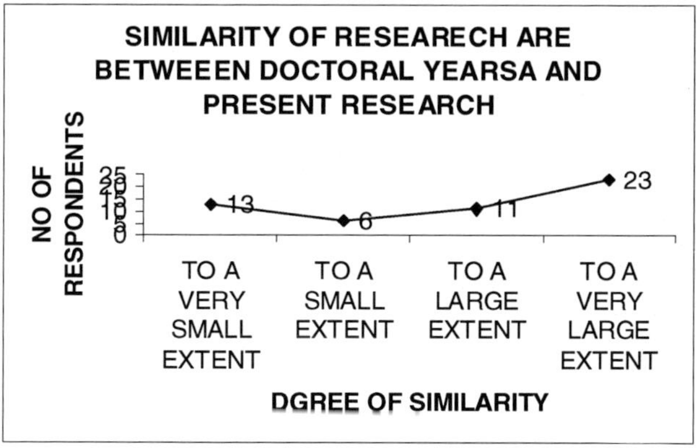

It is evident that 43.40% of the respondents have not changed their area of research to a very large extent. Another category of scientists (24.83%) of the sample reported the degree of similarity of research between the areas they began with, and the current area of research. The degree of similarity is only to a very small extent. It indicates that they were attracted to new areas of research that emerged during their professional career. (See Table.4.6.3).

Table.4.6.3: Extent of Shift of Research Area

DEGREE OF SIMILARITY	FREQUENCY	PERCENT
TO A VERY SMALL EXTENT	13	24.53
TO A SMALL EXTENT	6	11.32
TO A LARGE EXTENT	11	20.75
TO A VERY LARGE EXTENT	23	43.40
Total	**53**	**100.00**

The scientists who reported a large extent of similarity comprised 20.75% of sample whereas in the case of a very small proportion (11.32%), similarity to a small extent. On the whole we observed that that there are two categories of scientists: a) those who almost continued in the same area of research; and b) those who almost shifted away from the area or research they began with at the time of their research career. As mentioned above, the shift in research concerns is more noticeable among younger scientists. Perhaps the new areas open up opportunities for more original research and consequent recognition.

4.7 Information Technology and Reorganization of Communication Infrastructure

Information Technology has been playing a crucial role in reorganizing communication infrastructure in organizations concerned with teaching and/or research and extension. Information Technology enabled:

- Computer Mediated Communication
- Library Automation
- Laboratory Reorganization
- Access to worldwide information through Computers

The organizations, in which the respondents are located, show considerable variation in the availability of communication infrastructure, degree of access to it and the consequent reorganization. In this section we describe the Information Management System in scientific organizations as an adaptive variation from a simple non-networked library-laboratory information management system to that of a complex networked library-laboratory information management system.

4.8 Information Management System

Conventionally, the formal sources of knowledge for scientists have been journals and communication events such as seminars and conferences, which provided opportunities for both formal and informal interaction among scientists for exchange of ideas. In the recent times computer-mediated communication has become a significant source. We attempted to examine both types of resources, the library, the repository of journals and books, and the computer mediated communication through networking. We examined differences in the accessibility to these resources by looking at the physical arrangement at the level of laboratory of a research group in the organizations in which organic chemists in our study are located.

4.8.1. State Universities

We considered the laboratory as a techno-social structure comprising of technology, information and people as actors in the situation. Social interaction among researchers-scientist/teacher on the one hand and her/his student (doctoral, postdoctoral) collaborators in the laboratory is mediated by technology gadgets, equipment, chemicals etc., We also considered the status of the library as it seems to be an integral part of the information management system essential for production of knowledge and productivity of individual scientists.

Library

The state universities in the study have traditional libraries with print journals. The information in the library is managed in terms of manual cataloguing by cards for identification and manual retrieval.

Figure 4.8.1: Laboratory setting in a State University

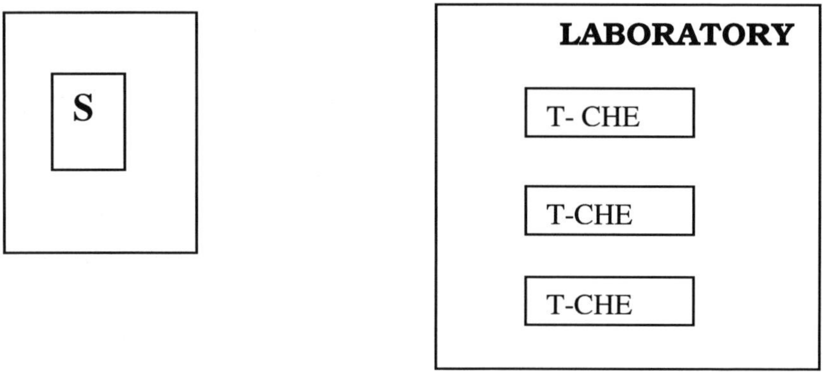

S-Stands for Scientist
CHE-Chemicals T-Table

The laboratories in the departments of organic chemistry in the state universities in our study have a typical structure with two separate rooms with the scientist accommodated in the room either adjacent to her/his laboratory or placed in some other location of the main academic building. The scientists and their collaborators have to move away from their room to either access the library or to go to the laboratory for the sake of experimentation. It also holds good for the research students. The scientists do not have any instrument of communication on their desk. They have to use a common facility situated in the department office. In most cases scientists did not have a computer. Only one scientist acquired a computer with project funds. He said that he uses his personal computer only for documentation work rather than for communication.

4.8.2 Central University

The information management system at the central university represents a quantum leap in all fronts compared to the state universities. The central university is way ahead in terms of the sources of information both in terms of quantity and quality. The library in the Central University has a good collection of the archives in the form of print journals and it subscribes to numerous reputed journals. The library is automated in the sense that one has access to on-line catalogue. The library also gives access to electronic versions of the major journals available via Internet campus wide network at the individual laboratories. At the time of the study complete digitization of the library was under way. In addition to this, we observed that individual laboratories have subscribed to some print and online journals, which are not acquired by the university library.

The Computer Center was established in 1985 as a central facility and to cater to the computational needs of the faculty, students and staff of the University. Barely four years into existence, the computer center recognized the importance of email and, with the help of NCST, Mumbai established an email connection through modem and a telephone line. In 1992, University Grants Commission funds allotted to the Computer Center helped to establish Thick/Thin Ethernet most of the campus in many improvised means for the faculty, students, and officers and other users.

In 1993, Educational Research Network (ERNET) a facility of the department of electronics (DOE), Government of India, selected the University to be one of the transit nodes with the addition of a VSAT dish antenna, but with a narrow bandwidth of 19 Kbps. In 2000, ERNET provided 128 Kbps microwave link. The Internet has expanded exponentially. The connection to the e-world has been enhanced to 2 Mbps in from ERNET in the year 2001.

The Laboratory Setting: The typical laboratory in the central university represents an important phase in the adaptive variation for access to knowledge and technology in the era of information revolution. The research groups exhibit a different organization of the work place. (See Fig.4.8.2)

Fig.4.8.2: Diagrammatic representation of the campus network

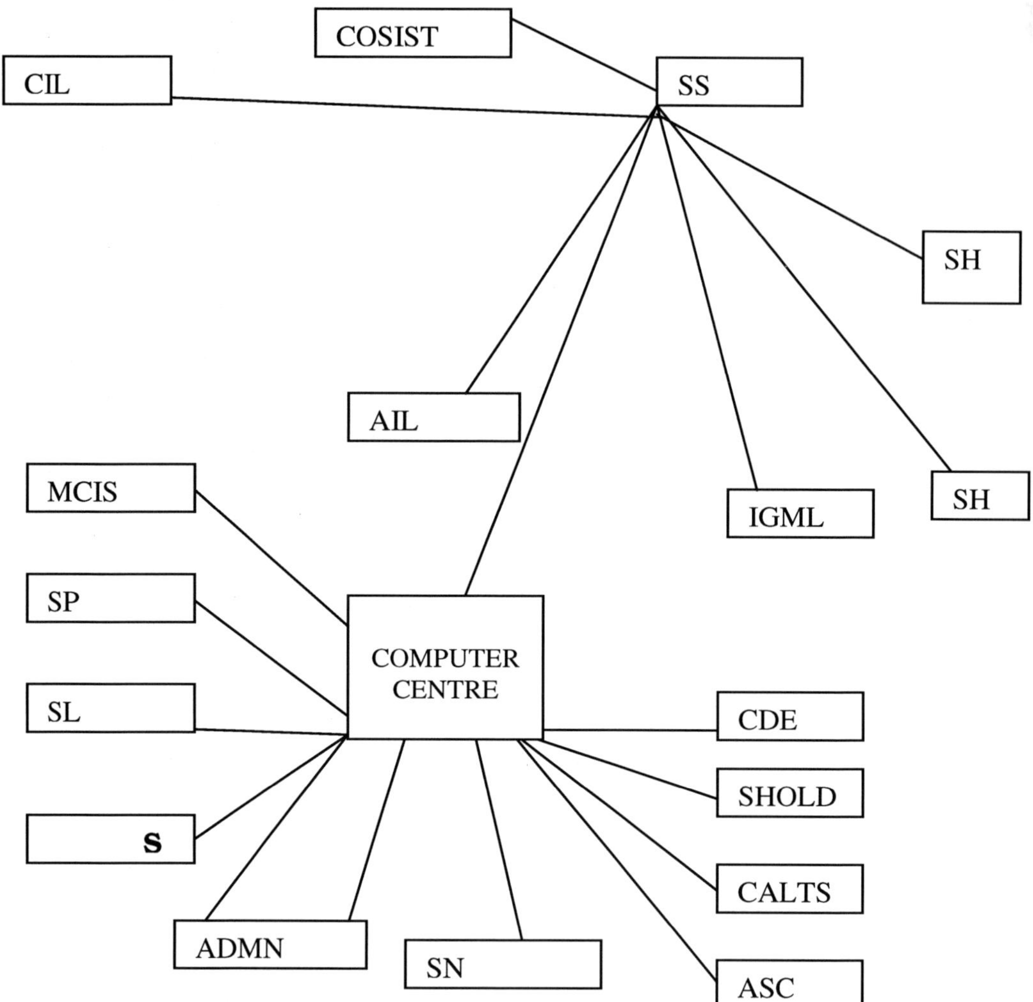

SC-Stands for School of Chemistry

Scientist situated in one room equipped with telephone and a computer is linked to Internet through the networked server located at the Computer Center. The laboratory, in addition to be a place of experimentation has emerged as a center for online learning as well. In addition to traditional workbenches, shelves of chemicals and infrastructure related to research, one notices a networked computer. (See Fig.4.8.3).

Fig.4.8.3: Typical laboratory setting at the central university

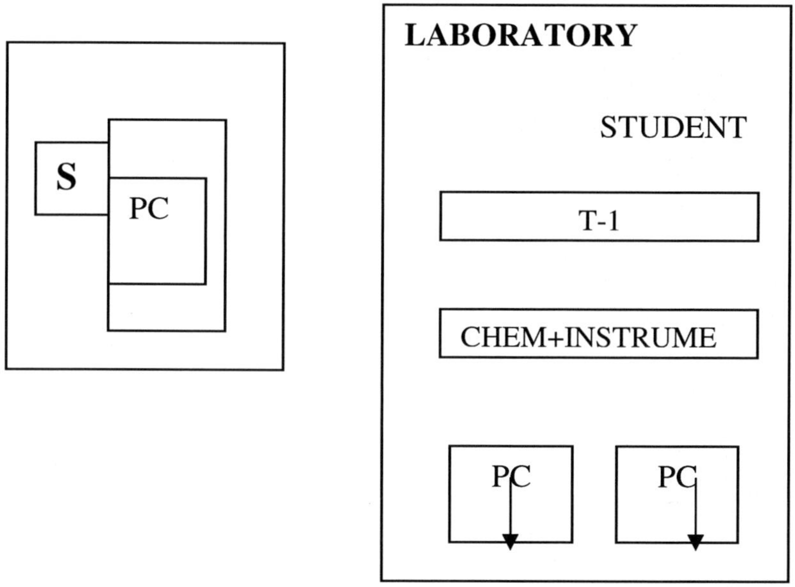

S-Scientists, **Stud**- Students, **CHEM**-Chemicals **Infra**-Infra-structure for experiment, **T**-Table **PC**-Personal Computer—Networked to E-mail and Internet Telephone

Thus the central university represents a new phase in evolution of laboratory setting which also accommodates the modern networked personal computer linked to the World Wide Web of knowledge.

4.8.3 The Indian Institutes of Technology(IITs)

At the IITs we observed that there is a distributed information management system where in the main server is connected to the proxy servers set up in the individual departments. The chemistry department runs its own server providing access to Internet and email to the individual scientists and the laboratories on the desktops. (See Figs.4.8.4 and 4.8.5)

Fig. 4.8.4.Campus Network in an IIT

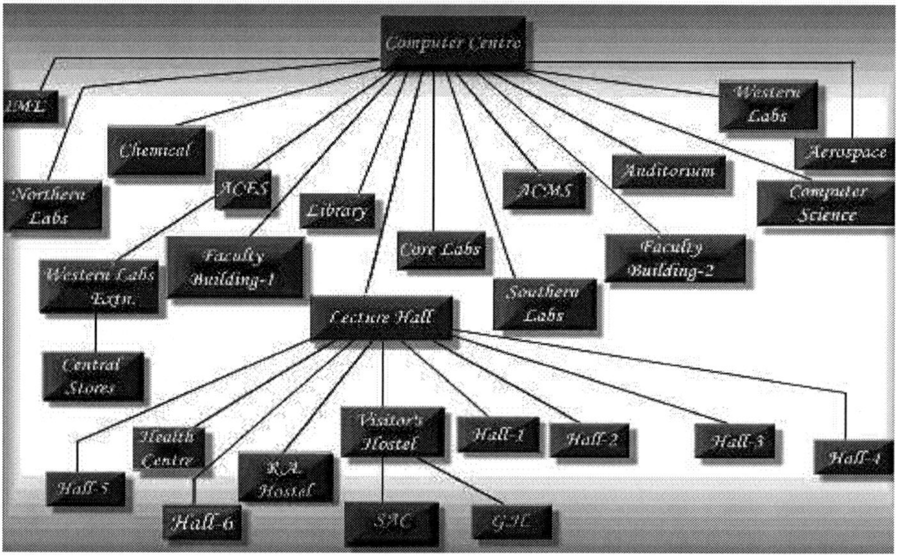

Fig 4.8.5: Campus network in an IIT

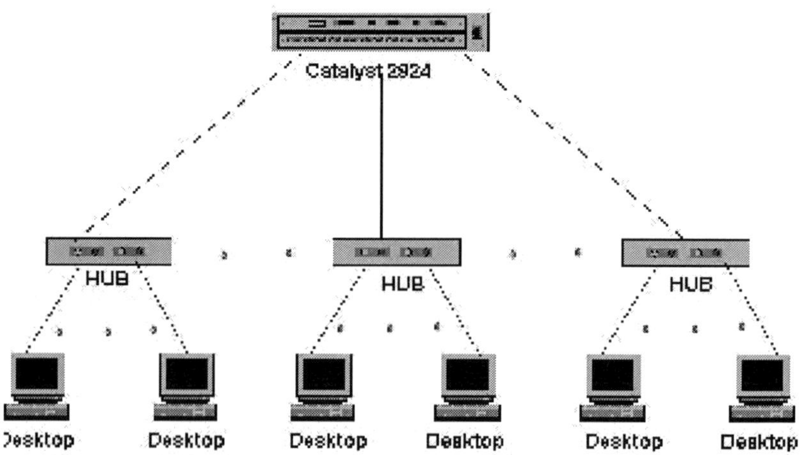

4.8.4 National Laboratories

The Library

At the national laboratory, the computerization of library services has been taken up. At present, Current Contents in physical, chemical and earth sciences, engineering, technology & computers and life sciences are obtained in print and diskette versions. Special CD-ROM services are available under CHEM BANK and PEST BANK.

Information Technology

As part of information technology modernization with world bank funds, the following intranet and internet infrastructure is being implemented at the national laboratory for implementing business process systems and electronic publishing (CD-ROM, WEB) for improved project monitoring and control.

Campus-wide Networking

Campus wide local area network provides data highway between twelve multistoried buildings across the Institute campus, which are located within half-kilometer radius. This system provides100 Mbps speed for backbone, network servers and power workstations and 10 Mbps for normal desktops. The system is designed for 250 nodes. It is a heterogeneous networking system and operates basically under Windows NT, with other environments such as Novell NetWare and SCO/SGI/DEC UNIX and network managers SNMP, Intranet WAN manager. Over 2-km fiber and 15 km. UTP have been used for establishing campus connectivity. In addition the campus is widely networked through Intranet, Internet systems and closed circuit-TV System. (See Fig 4.8.6)

Fig..4.8.6: Information Network Layout in the National Laboratory

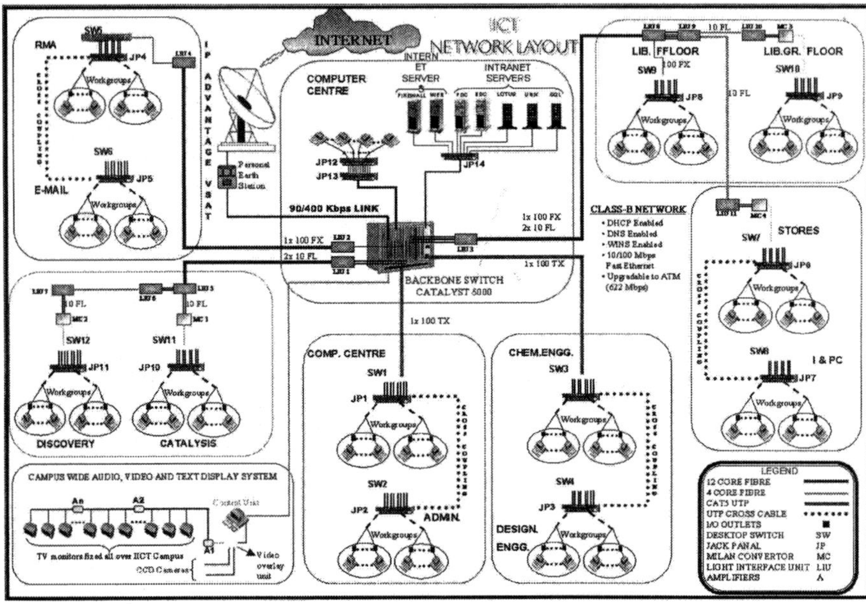

4.8.5 The Deemed University

The information management system in the deemed university is structured in a hierarchical manner in terms of print journals and online information resources. The library contains all reputed journals. The individual departments have their own collection of journals. The library is networked to campus computer center and completely digitized. (See Fig.4.8.7.1)

Fig.4.8.7.1: Information Management System –Deemed University

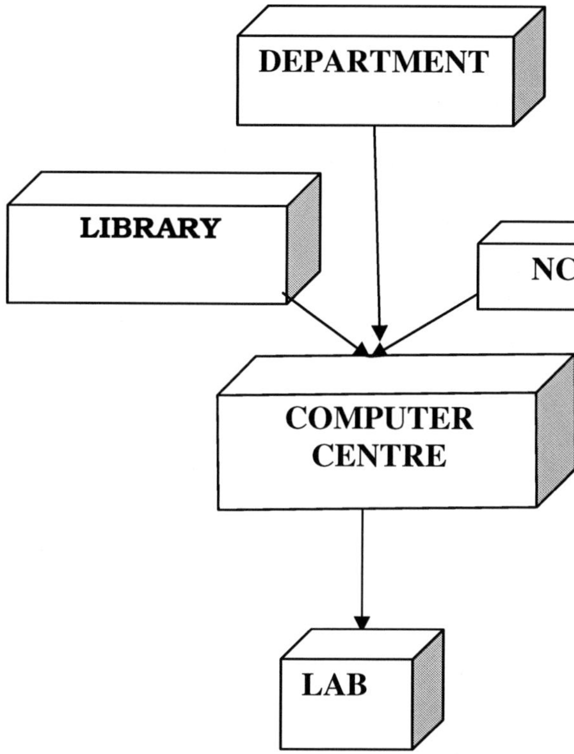

* The cube represents a laboratory connected to library as well as computer centre with its own proxy server.

4.8.7.2 The Laboratory Setting

A typical laboratory in the deemed university is a modernized version of a traditional chemistry laboratory .The laboratory in the center accommodates rows of tables with chemicals and infrastructure. On either side of the table we find stacks of reprints of research publications on one side and a minimum of three computers networked to World Wide Web and email. We observed that in the deemed university, the researchers have instant online access to information next to the research table.

Thus the information management system at the deemed university represents advanced, complex and networked model for more user-friendly purpose to the researchers providing parallel access to information from various sources while the experiment is on.

Fig.4.8.7.2: The laboratory setting-deemed university

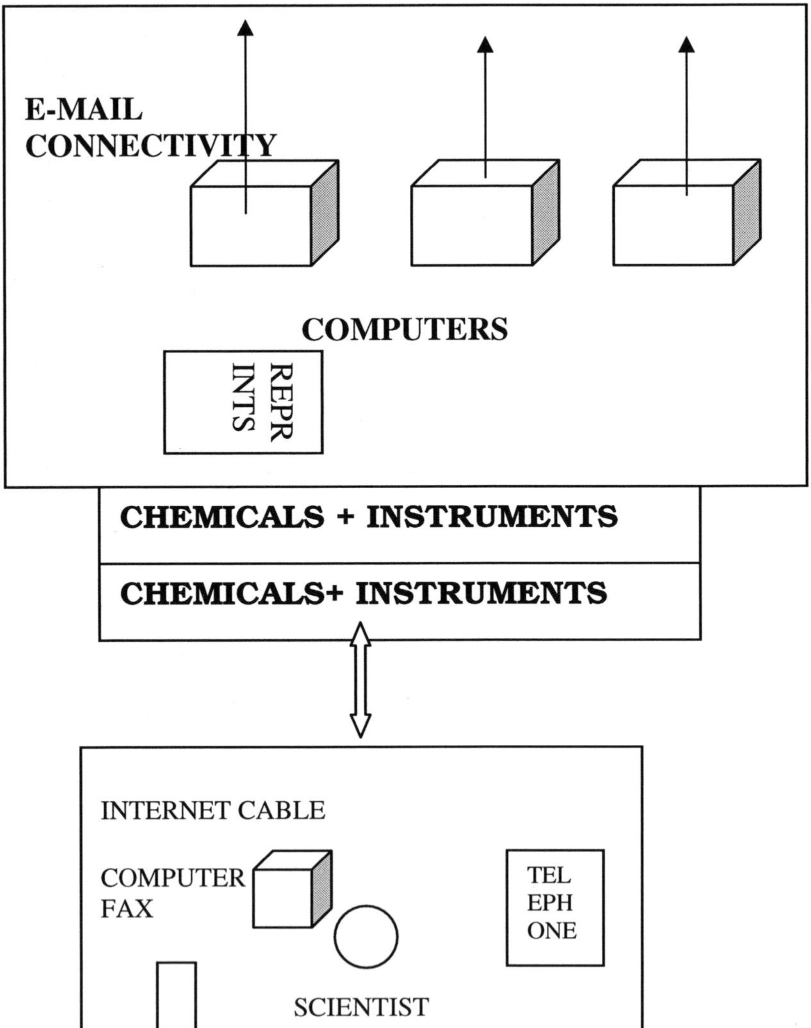

4.9 Access to Communication Technology

Different media of communication facilitate communication in modern science. In our study, we assessed the relative access, mode of procurement and frequency of use of these media in communicating ideas by the scientists from different organizations. We describe the level of access in the first place followed by the mode of procurement of these instruments. (See Table.4.9)

All the respondents from Central University, IITs, national laboratories, and deemed university have access to telephone. They have telephones on theirs desks and only 63% of the scientists from the state universities have access to telephone.

Respondents from central university, IITs, and deemed university have access to FAX either at the individual laboratory, department or school level. Only 36% from state universities and 71% from the national laboratories have fax facility that is easily accessible.

All the respondents from central universities, IITs, and deemed university and 85% of respondents from national laboratories have access to computers. They have computers either on their desks or in their labs. In contrast, in the case of State universities, only 45% of the respondents have computers on their desks or in their labs.

Table.4.9: Access to instruments of communication- organization wise

INSTRUMENT	ORGANIZATION	YES	NO	TOTAL
TELEPHONE	STATE UNIVERSITY	7	4	11
	CENTRAL UNIVERSITY	10	0	10
	IITs	13	0	13
	CSIR LABS	14	0	14
	DEEMED UNIVERSITY	5	0	5
	Total	49	4	**53**
FAX	STATE UNIVERSITY	4	7	11
	CENTRAL UNVERSITY	10	0	10
	IITs	13	0	13
	CSIR LABS	10	4	14
	DEEMED UNIVERSITY	5	0	5
	Total	42	11	**53**
COMPUTER	STATE UNIVERSITY	5	6	11
	CENTRAL UNVIERSITY	10	0	10
	IITs	13	0	13
	CSIR LABS	12	2	14
	DEEMED UNIVERSITY	5	0	5
	Total	45	8	**53**
EMAIL AND INTERNET	STATE UNIVERSITY	5	6	11
	CENTRAL UNVIERSITY	10	0	10
	IITs	13	0	13
	CSIR LABS	13	1	14
	DEEMED UNIVERSITY	5	0	5
	Total	46	7	**53**

Access to e-mail and Internet

All the respondents from central university, IITs, deemed university and 92.85% from national laboratories have access to e-mail and Internet. They access e-mail and Internet through their computers on their desks. In case of state universities, only 45% have access, that too not at the institutional level. They access these facilities either from their home PCs or private cyber cafes.

4.10 Mode of Procurement of Communication Gadgets

In the study we identified the sources from which they procured the communication equipment for their scientific research (See Table.4.10)

In our study, 42 out of 53 respondents acquired telephone with project funds, followed by 3 scientists who acquired the connection as part of in-house support. Out of the eight respondents who do not have access to telephone, four of the scientists are from the state universities. Thirty-two respondents have acquired FAX machines as a central facility. Twenty-one respondents purchased FAX machines from project funds. On the aggregate, 9 respondents have reported the absence of FAX machines.

In case of State universities, out of 11 scientists, only 4 have personal computers in their offices or labs. Of them 1 acquired the computer through in-house support and 3 from project funds. In the case of the central university, only one scientist acquired computer from in-house funds and the rest from their project funds.

Majority of the respondents from IITs, national laboratories, acquired personal computers through institutional resources. In case of the deemed university only one scientist acquired personal computer from in-house funds and the other 4 acquired personal computers from their project funds.

Thus it appears that communication technology has come to occupy a key role in scientific research in Indian science too, with almost all the scientists from different organizations recognizing the importance of these media, and procuring them for research communication. It also appears that, at present, communication technology is acquired only through extra-organizational sources like project funds.

Table 4.10: Mode of procurement of communication technology.

INSTRUMENT	ORGANIZATION	INSTITUTION	PROJECT FUNDS	ANYOTHER	ABSENT	TOTAL
TELEPHONE	STATE UNIVERSITY	0	7	0	4	11
	CENTRAL UNIVERSITY	0	6	0	4	10
	IITS	2	11	0	0	13
	CSIR LABS	1	13	0	0	14
	DEEMED UNIVERSITY	0	5	0	0	5
	TOTAL	**3**	**42**	**0**	**8**	**53**
FAX	STATE UNIVERSITY	2	2	0	7	11
	CENTRAL UNIVERSITY	1	2	0	7	10
	IITS	2	4	0	7	13
	CSIR LABS	3	9	0	2	14
	DEEMED UNIVERSITY	1	4	0	0	5
	TOTAL	**9**	**21**	**0**	**23**	**53**
COMPUTERS	STATE UNIVERSITY	1	3	0	7	11
	CENTRAL UNIVERSITY	1	3	0	6	10
	IITS	9	4	0	0	13
	CSIR LABS	11	2	0	1	14
	DEEMED UNIVERSITY	1	4	0	0	5
	TOTAL	**23**	**16**	**0**	**14**	**53**

4.11 Source of Funding for E-Mail & Internet

E-mail and Internet facilities ushered in a revolution in scientific communication. For example, communicating manuscripts for publication, and the consequent correspondence with the journals are done through e-mail. Further, announcements regarding conferences, seminars and fellowships are made on the Internet. PC with Internet facility has become an indispensable tool for communication apart from its uses for data and text processing.

I tried to identify to what extent the organization supports the scientists in acquiring or gaining access to communication facilities like e-maim and Internet. These are relatively new and efficient modes of communication, which tend to influence the productivity of scientists. We observed that excepting the state universities, all the other organizations have provided access to the email and Internet. (See Table.4.11)

Table.4.11: Source of funding for e-mail & internet

ORGANIZATION	INSTITUITION	PROJECT FUNDS	ANY OTHER	ABSENT	TOTAL
STATE UNIVERSITY	1	3	3	4	11
CENTRAL UNIVERSITY	10	0	0	0	10
IITS	13	0	0	0	13
CSIR LABS	14	0	0	0	14
DEEMED UNIVERSITY	5	0	0	0	5
TOTAL	43	3	3	4	**53**

Four scientists reported the absence of access to email and Internet. We also found that 3 scientists arranged access to email and Internet through project funds and other means like nearby cybercafes, or browsing and mailing from home.

Thus it appears that organic chemists in our study have recognized the role of communication technology in the production of knowledge and seem to gain access to the technology through various ways.

4.12. Extent of Use of Email

The scientists were asked to indicate as to how long they have been using e-mail facility for purposes of research communication.

Table 4.12: Extent of use of email

ORGANISATION	1YEAR	1-2 YEARS	3-4 YEARS	5+YEARS	NO USE	TOTAL
STATE UNIVE	2	2	5	0	2	11
CENTRAL UNIV	0	0	3	7	0	10
IITS	0	3	5	5	0	13
CSIR LABS	1	4	6	3	0	14
DEEMED UNIV	0	0	0	5	0	5
TOTAL	3	9	19	20	2	**53**

Majority of the scientists (39) mentioned that they have been using the e-mail facility for the last 3 to 5 years. In this category, we observed that all the scientists are from central and deemed universities and 50% of the scientists from IITs and national laboratories belong to this category. This is followed by 9 scientists who started using e-mail communication facility between 1-2 years ago and 3 scientists for less than an year, of which 2 are from the state university and one more from a national laboratory. Two scientists from state universities do not use e-mail facility at all.(See.Table.4.12)

4.13. Number of Hours of Browsing on the Internet in a day by the Research Group

Print journals, continue to be one of the sources of ideas for scientists. Recent developments in computer network technology enabled the possibility of creating electronic journals and different search-engines. Through these scientists can get information on their desks on the World Wide Web.

It is observed that scientists in addition to going to the library also find literature on the web. Through the questionnaire we supplied and also from the in-depth interviews we obtained information from each research group, on the number of hours in a day the members of the group browse on the net.

Table.4.13: Number of hours of browsing on the internet in a day by the research group

ORGANISATION	0HRS	<1 HOUR	< 2HOURS	<3HOURS	<4HOURS	<5HOURS	>5 HOURS	TOTAL
STATE UNIVE	6	3	1	0	0	0	1	11
CENTRAL UNIV	1	6	3	0	0	0	0	10
IITS	1	3	4	3	0	1	1	13
CSIR LABS	5	3	4	1	1	0	0	14
DEEMED UNIV	0	0	2	2	1			5
TOTAL	13	15	14	6	2	1	2	**53**

We considered the average of the number of hours, browsed on the net related to research, both by the scientist and the members of the research group. We categorized the duration of browsing in terms of number of hours. There are thirteen scientists in the category of those who do not browse at all on the web for research related information. Of the 13, 6 are from state universities followed by 5 from national laboratories and one each from central university and IITs. We observed that, the majority of research groups 29 out of 53, reported that they browse between ½ an hour to 2 hrs a day. Research groups further varied as, as 6 groups reported between 2-3 hours and 5 groups above 3 hours of browsing in a day.

4.14. Collaboration

In the earlier section, we have seen the way the Information Technology reorganized infrastructure for scientific communication. Computers increased access to peer group and literature. For collaboration, communication is a prerequisite.

Modern science is not a single man's pursuit. Science as a social institution necessitates interdependence among the practitioners in a given research arena. Sociological research offers many reasons and explanations for the question - why scientists collaborate. We aimed at understanding why organic chemists in the study collaborate with others? If so, how this cooperative attribute varies according to organizations in which scientists are located especially in the Indian context. Answers to these questions would reflect the character of the community of organic chemists. We inquired into how many among our respondents collaborate, with whom do they collaborate and the institutional affiliation, of the collaborators.

Table 4.14.1: Status of Collaboration

ORGANIZATION	NON-COLLABARATORS	COLLABORATORS	TOTAL
STATE UNIVERSITY	7	4	11
CENTRAL UNIVERSITY	3	7	10
IITS	2	11	13
NATIONAL LABS	2	12	14
DEEMED UNIVERSITY	0	5	5
TOTAL	**14**	**39**	**53**

In the State universities more than half the total number of respondents were not engaged in collaborative research. Seventy percent of organic chemists from central university collaborate. Among those from IITs, National laboratories, 90% of the respondents mention that they are engaged in collaborative research. Whereas, all the scientists from the deemed university are actively involved in collaborative research. (See Table.4.14.1)

4.14.2 Number of Collaborative Projects

In our study, after identifying scientists who were engaged in collaborative research and those who were not, at the time of the study, we categorized them into 5 groups namely those with no collaboration, those with 1 collaborative project, followed by those with 2, and 3 and 4 or more than 4 collaborative projects. The distribution is as follows.

I observed at the time of the study, that 14 respondents out of 53 were not involved in collaborative research. Fifty percent of those who did not collaborate hail from the state universities. All the scientists from the deemed university are involved in collaborative research and have more than 2 collaborative projects. A total number of 11 respondents have either 4 or more than 4 collaborative projects, followed by 11 with 1 collaborative project each, 10 with 3 collaborative projects each and 7 scientists with 2 collaborative projects each. (See Table.4.14.2)

In case of the central university, 3 scientists are involved in either 4 or more number of collaborative projects, followed by 2 with 1 collaborative project and 1 each with 2 and 3 collaborative projects. In the case of the IITs, 4 scientists have either 4 or more number of collaborative projects, followed by 3 scientists with 1 collaboration each, 2 scientists with 3 collaborative projects each and one scientist with 2 collaborative projects. In case of the national laboratories, only one scientist has around 4 collaborative projects and 4 scientists each have 1, 2 and 3 collaborative projects each.

Table 4.14.2: Number of collaborative projects

ORGANIZATION	0	1	2	3	4 and above	TOTAL
STATE UNIVERSITIES	7	2	1	1	-	11
CENTRAL UNIVERSITY	3	2	1	1	3	10
IITS	2	3	1	2	4	13
CSIR LABORATORIES	2	4	4	4	1	14
DEEMED UNIVERSITY	-	-	-	2	3	5
TOTAL	14	11	7	10	11	**53**

It may be said that the scientists trained at reputed institutions and located at well-endowed research institutions appear to see the significance of collaboration. Scientists from state universities seem to insulate themselves.

4.14.3. Intra-institutional (within the country)

It is observed that 37.93% of our respondents reported that they are involved in collaborative work with their colleagues in the same organization. In the case of organic chemists from the state universities, a very small numbers of respondents were found to collaborate with people within the organization.

Table.4.14.3: Intra-institutional collaboration (within the country)

ORGANIZATION	NO OF RESPONDENTS	TOTAL	PERCENTAGE
STATE UNIVERSTIY	4	11	36.66%
CENTRAL UNIVERSITY	3	10	27.27%
IITS	3	13	23.07%
NATIONAL LABS	6	14	42.85%
DEEMED UNIVERSITY	4	5	80.00%
TOTAL	**20**	**53**	**37.73%**

In this respect, 80% of respondents from deemed university, collaborate with colleagues within and outside the organization. In case of national laboratories, 42.85% of the respondents are collaborators, followed by the central university and IITs with 25 percent of respondents. (See Table. 4.14.3)

4.14.4. Inter-Institutional Collaboration (within the country)

The study aimed at knowing in the extent of collaboration between researchers in the study on the one hand and those from the other research organizations like academia and industry on the other.

Majority of the respondents from State Universities seem to collaborate only with scientists from state universities and not with those from any other institutions. Only one scientist from a state university has been engaged in collaborative research with the industry.

Table.4.14.4: Inter-institutional collaboration (within India)

ORGANIZATION	STATE UNIVERSITY	CENTRAL UNIVERSITY	IITS	CSIR LABS	DEEMED UNIVERSITY	INDUSTRY
STATE UNIVERSTIES	4	-	-	-	-	1
CENTRAL UNIVERSITY	-	3	-	2	3	-
IITs	-	-	3	-	-	8
NATIONAL LABS	-	1	-	6	1	5
DEEMED UNIVERSITY	-	1	-	1	4	3

Majority of the scientists from IITs (8) reported that they collaborated with industry. (See Table 4.14.4). Only 3 had collaboration with scientists from other IITs. Scientists from national laboratories seem to have extensive collaboration with those from other national laboratories. From the deemed university 4 scientists were involved in collaboration with present colleagues, while one scientist each collaborated with one each from central university and national laboratories. Also three of the scientists have collaborative projects with industry.

4.14.5. Inter Institutional Collaboration (Outside the Country)

Collaborative research with scientists from other countries has always been an important aspect of the visibility of the scientists. In the present study we attempted to find out how many of the scientists collaborate with scientists from abroad, and how does this vary across different scientific organizations in India.

We identified three different kinds of foreign settings in which the Indian scientists collaborated with their counterparts. These settings are foreign universities, foreign institutes, and foreign industries.

Fig.4.14.5: Pattern of International Collaboration

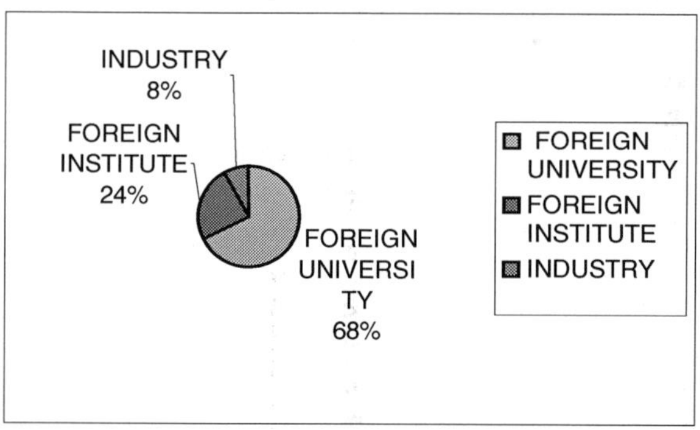

Scientists collaborate more (68%) with their counterparts in foreign universities, followed by those in the foreign research institutes. A significantly small number (8%) of scientists collaborate with foreign industry. (See Fig.4.14.5)

Scientists from deemed universities, central universities and national laboratories, collaborate with their peers' foreign universities whereas among those from state universities none reported such collaborative work. Excepting the scientist from IITs, scientists from all other types of organizations collaborate with those in foreign institutes. In this category only one scientist is from a State University. (See Table.4.14.5)

Table.4.14.5: Pattern of International Collaboration

ORGANIZATION	FOREIGN UNIVERSITY	FOREIGN INSTITUTE	INDUSTRY
STATE UNIVERSITY	-	1	-
CENTRAL UNIVERSITY	7	4	1
IITS	8	-	1
NATIONAL LABS	4	4	2
DEEMED UNIVERSITY	4	3	-

Excepting those from both state university and deemed university, scientists from other organizations like central university, IITs, and national laboratories collaborate with foreign industry too.

4.15. Productivity Differentials among Respondents

Social scientists of science, for long have been debating over universal indicators to evaluate the performance of the scientists. Researchers have been using the criteria of number of publications, citations and journal impact factors to comment on the quantity and quality of publications.

In our study, we collected data from scientists regarding their publications. We realize that it is important to realize that from a phenomenological perspective, the actors' (scientists) notion of What is productivity is, appropriate measures of productivity and how the context, (in this study their situatedness in India influences productivity what purposes does it serve. Organic chemists expressed their perceptions regarding the quantity and quality aspects of productivity.

In this study, productivity was defined as the number of research publications of the scientists that were published in refereed journals. We selected a five-year period (1996-2000) for the purpose of counting the publications, as this would give a stable picture of publication pattern. We organized the number of publications into classes with an interval of 10 each. (See Table.4.15)

Scientists from the deemed university have to their credit 30 and above number of publications. In the sample, 15 respondents published between 20-29 articles. In case of state universities, excepting one scientist who published above 40, all others fall in the category 1-29 publications. In case of the central university, the scientists have publications falling in the category of 20-50 and above.

Table.4.15: Productivity Levels Organization Wise

	TYPE OF ORGANIZATION					
PRODUCTIVITY	STATE UNIVERSITIES	CENTRAL UNIVERSITY	IITS	CSIR LABS	DEEMED UNIVERSITY	Total
0	-	-	-	3	-	3
1-9	4	-	2	3	-	9
10-19	2	-	2	2	-	6
20-29	4	5	5	1	-	15
30-39	-	2	2	2	2	8
40-49	1	2	2	1	1	7
50-59	-	1	0	2	2	5
Total	**11**	**10**	**13**	**14**	**5**	**53**

The Scientists at IITs, majority of who were young scientists, fall in the category of 1-49. The CSIR laboratories, which have exclusive applied divisions, show the maximum variability in terms of the productivity levels. Two scientists from the natural products division of a national laboratory, belonging to organic synthesis, who have been collaborating with each other and scientists from other organizations within and outside India, have published more than 50 papers each. Another 6 scientists who published between 20-49, and followed by 5 scientists have published below 19.

4.16 Comparison of Averages of Publications-Organizationwise

It is relevant at this stage to compare the average number of publications during the years (1996-2000) from all the types of organizations. (See Table.4.16). It is noticeable that the deemed university leads in terms of averages of publications followed by the central university. The national laboratories occupy the third position followed by the IITs. The state universities are far behind all the other types of organizations. We calculated the average number of publications during 1996-2000 by dividing the total number of publications by the number of respondents per type of organization in the study.

Table.4.16: Comparison of averages of publication productivity

SL.NO	ORGANIZATION	TOTAL NO.OF PUBLICATIONS	AVERAGE (1996-2000)
1	STATE UNIVERSITIES	180	16.36
2	CENTRAL UNIVERSITIES	358	35.8
3	IITs	331	25.46
4	NATIONAL LABS	367	26.21
5	DEEMED UNIVERSITY	257	51.4

From the data (See Fig.4.16) it appears that productivity levels among organic chemists are basically polarized in terms of two groups. The Deemed University, The Central University, The National Laboratories, and the IITs represent the elite in terms of productivity levels with the state universities at the other extreme.

Fig.4.16: Comparison of averages of publication productivity

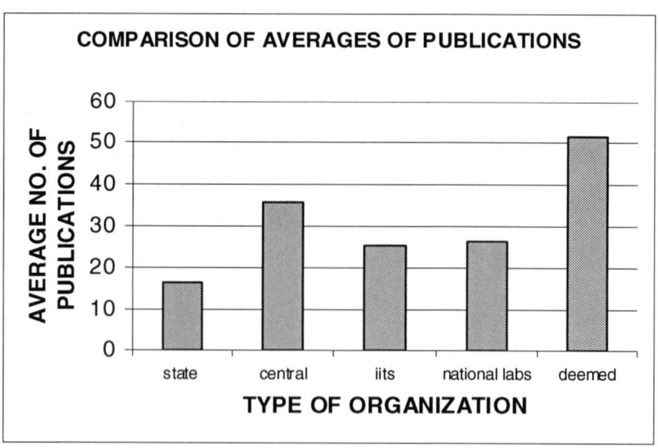

4.17 Summary of Findings

In this chapter, the profile of the organic chemists included in the study is preented. In the first place we studied the composition of the scientists according to the age groups. We found that the youngest respondent in the study is 28 years and the oldest is 59 years old with average age being 46.07years.

We studied the current organizational affiliation of the scientists and related it to the organization the scientists pursued their doctoral degree at. We found that scientific organizations like the Deemed university, Central University, IITs, and National Laboratories, recruit scientists who obtained doctoral degree from similar organizations but not from the state universities. We observed that scientists who were recruited into the state universities were trained at the state universities only.

The study shows that postdoctoral training has become an essential precondition for recruitment in scientific organizations in India. While scientists at the deemed university, the central university, the IITs represent a particular average number of years of postdoctoral training especially outside India, the state universities do not seem to attach much significance to doctoral training in India or abroad. In case of national laboratory, the scientists belonging to the younger age groups seem to have undergone postdoctoral training while the ones above 49 years never underwent postdoctoral training.

We studied the economic resources for research activities of organic chemists in the study. We found that, the amount of in-house budget for recurring expenditures varies from one organization to another as they differ in terms of their resource endowment levels. We also observed that scientists receive not more than 25% of the non-recurring expenditure from in-house resources. The in-house financial resources over the past 5 years have remained the same. We found that scientists have been mobilizing the rest of the resources through alternative means like research projects sponsored by pubic and private agencies, and taking up industrial consultancy projects. Nearly 72% of the respondents undertake sponsored projects from different sources like DST, CSIR, DBT and DAE. Regarding consultancy projects, only less than 50% of the respondents undertake consultancy projects. Thus the study reveals that disproportionately more resources are allocated to a few institutions where there is a concentration of recognized scientists. In our study we find that, there is a gradation of institutions in terms of allocation of resources. The Deemed university, the IITs, the national laboratories and Central University occupy prime positions, where as, the state universities occupy the back seat.

We observed that in terms of specialization in organic chemistry, 46% of the respondents specialize in organic synthesis followed by 23% in bio-organic chemistry. 7 scientists in natural products, 6 in supra-molecular chemistry followed by 4 scientists in Organo-metallic chemistry.

In terms of opting for specialization, we observe that in the case of state universities, scientists seem to specialize in more traditional areas of research such as natural products and organic synthesis. Whereas scientists from other organizations in the study are spread over all specialties of organic chemistry. The study shows that, specialties like organic synthesis and natural products are the domains of older scientists. Scientists in the age group of 40-49 represent majority of each sub-discipline. We see that younger scientists and a few older scientists are attracted to frontier areas of research in the new and emerging specialties.

We explored the similarity of current research area of organic chemists with that during the doctoral years. We find two diametrically opposite trends. Scientists continue in similar area of research or completely shift to another area of research.

We attempted to understand the pattern of information management system in each type of organization in the study. We observed that in Indian scientific organizations, communication technology reshaped the information management system Now there is a greater connectivity between library-laboratory in reputed, well endowed institutions like the deemed university and the IITs. In the case of state universities, the library and laboratory are not only poorly equipped but not connected as well. Whereas in the case of other organizations, computer mediated communication system emerged as an integral part of both library and laboratory settings and also integrated them.

We observed that technology related to communication in science like telephone, fax and computers are omnipresent among scientists from all the organizations except the state universities. More than 50% of the scientists from the state universities have no access to such instruments. In most of the cases such technology is acquired from the project funds. and maintenance is by the institution in case of first three organizations, but in case of state universities are do not have and in the cases of those access from private agencies and collaborators. We also found that the extent of use of email facility for communication purpose is aged over 5 years in elite organizations where as very recent and yet to begin in the state universities.

The findings suggest that among the scientists who have been engaged in collaborative research 90% are from the elite organizations, whereas less than 50% of the scientists from state universities collaborate. Fifty percent of those who did not collaborate hail from the state universities. All the scientists from the deemed university have been found to be involved in collaborative research. Intra-organizational collaboration is specific among state universities and deemed university. As far as the collaboration with industry is concerned, scientists from the IITs collaborate more with industry followed by those from the deemed university and least number of scientists from state universities. Sixty eight percent of the scientists collaborate more with those in the foreign universities, followed by those in the foreign research institutes. A significantly small number (8%) of scientists collaborate with foreign industry. Scientists from deemed university, central university and national laboratories, collaborate with those from foreign universities whereas among those from state universities, none reported such collaborative work.

We found that scientists from the Deemed University, Central University, National Laboratories, and the IITs are more productive in terms of the number of publications compared to scientists at the state universities.

In the next chapter, we shall examine what factors are responsible for productivity differentials among organic chemists in India and analyze their implications for Science in India.

CHAPTER FIVE

FACTORS INFLUENCING PRODUCTIVITY

Introduction

In this chapter, I attempt to explain the productivity differentials among organic chemists in the Indian context. I shall examine the relative contribution of factors I considered in the study, to the productivity levels and differentials among organic chemists in India. This chapter is divided into two sections.

In the first section, I focus on productivity in Indian science in terms of how organic chemists perceive productivity in the Indian context, and what factors, according to them account the productivity of a scientist. I discuss and elaborate upon the quantity and quality aspects of productivity.

In the second section, I attempt to explain the relationship between the productivity levels and organizational factors like doctoral affiliation, postdoctoral training, duration of association with the current organization of employment, the demographic variable, age of the respondents and also to the similarity of research area.

5.1 Productivity

Social scientists proposed different measures to evaluate the performance of scientists and proposed indices on the basis of the measures. Productivity is considered in terms of number of publications, a quantitative measure, and other measures like, citation index and impact factor analysis.

In the first place, I examine what is meant by productivity for organic chemists in the Indian context and what factors contribute to the productivity of organic chemists. I examine the debate on the difference between quantity and quality aspects of productivity. I also examine what role does productivity play in different professional aspects of a scientist's career.

Responding to the question on what productivity meant to each of the scientists in the study, during the course of the interviews, the scientists revealed the following meanings of productivity.

5.1.1Number of Publications as a measure of Productivity

Many respondents emphasized upon the number of publications as a quantitative measure of productivity. Some respondents mentioned about indicators like citation index and journal impact factors as measures of quality. However, many scientists mentioned that, consideration of citation index and journal impact factor as measures of quality of performance is only beginning to take roots.

In the course of interviews, some respondents emphasized upon publications an art of survival in science. They said that in science, one has to continuously publish otherwise one cannot survive. To quote the response of a professor of organic chemistry from IIT to the question

> "How is the productivity of a scientist measured?
>
> I don't know. In academics as far as I know it is still 'publish or perish'. Okay, in that there are no changes, even though there are folks saying that they have published and got patents and prospered. That is what Mashelkar says.
>
> But we are still rated in publications, there are people who do lot of applied work and patenting. But that number is still very small percentage. So mostly it is published."

But when asked about the quantity and quality of the journal, says a scientist from IIT:

> "Well there is a slightly different way of answering this there are, especially, even for recruitment or publication.
>
> For recruitment also we have some norms. We call them norms. So we say that after . you should have at least 3 years and they also should have 'n' number of publications. So 10 or 12 or some publications i. e. they in the administration have certain norms. I would imagine somebody, may be like registrar or deputy registrar. They are purely administrative and they have looked at it over a period of time and they have said okay. These are the norms etc,, I am talking about it in our institution not other institutes, we differ from institute to institute, so I know personally there have been instances, when candidates apply, like you have said, the quality matters or

the number matters , In quite a few instances, it looked like the number came. Because there could be a guy who would have published in some kind of mediocre journal, 10 and 12 publications and another guy would have published in reputed journals like in chemistry you have the journal of American Chemical Society,, Tetrahedron, or Nature, the guy could have one Nature publication which could outweigh 15 other publications, pardon me for saying like this , Indian journal of chemistry is okay.. But on paper, he is still substandard he doesn't meet the quality. He doesn't meet the standards, so this has happened. I am not talking about somebody publishing in NATURE. I am talking about someone who has 5 good publications in some ACS journals and Royal society journals, another guy having in obscure journals some 15 publications. We have seen papers came back saying " why did u short-list this guy? He does not at all meet the publication NORMS. And still we have a tough time overcoming this.. But there should be some way you should evolve a formula may be including the impact factors of the journals and so on which exist right now. May be you can multiply that and get some value. You know, things like that, May be if you look at it some more scientifically then it is okay. But as of now, it is only number, pure number."

For purposes like recruitment and promotion, IITs emphasize upon some minimum number of publications. Scientists from the IITs say that the administration demands that the applicants must have a specific number of publications. They are very clear about the fact that, the administration only emphasizes upon the numerical count of publications. Some respondents quoted the instances of rejection of the candidature in case of applicants with less number of publications in high impact factor journals whereas the administration considered cases of organic chemists who published a comparatively higher number of research publications though in low impact factor journals. The authorities in charge of evolving norms still seem to operate the norm of quantity i.e., number rather than the quality i.e. publications in high impact factor journals. Thus they say that at this point of time, only the number counts. In such a situation, the scientist form IIIT says that it must be left to knowledge peers rather than administrators.

Some respondents mentioned that productivity is a difficult thing to define because in case of new and emerging areas of research, the number of people engaged in research is small and thus the area of research in certain cases becomes a limitation to define productivity. In other words, productivity depends on the scope of cognitive development of a research field. One scientist remarked:

"I don't know you are asking a wrong person. Well the most common way of doing it is... to have a formula, number of papers he has published multiplied by how much he has published multiplied by how many people have quoted it divided.. How many rupees have been spent. That is the formula...That is a very mundane formula of course, in the sense of any other yardstick, People use it for certain things. I think it's not true. That is one of the factors which people use to vary one over the other. Those things don't come into picture regarding funding or supporting some kind of research. Some research if nobody is doing if only you are doing, obviously nobody is going to quote you. So productivity is a very difficult thing to define in science. Views differ from people to people. Here we have Prof. Srinivasan who says few papers in his field are fantastic. Whereas the field in which I work, it should be more like 5 to 6 Not that there is any scientific purpose but people of my age, my caliber those who have enough money will publish that much."

5.1.2 Number and Career

The majority of the scientists said that, the existing evaluation system in Indian science insists on the number of publications. One respondent mentioned that producing number is not difficult but insists that productivity of a scientist must be seen in terms of overall contribution to science that could be explained not by numbers of publications alone. The same scientist also said that one has to publish in journals of higher impact factor and must be quoted by many instead of publishing more number of publications in low impact factor journals.

To the question, the excerpt from an interview with one of the scientists reflects whether the number of publications alone counts for a promotion or an award- the typical response:

"Yeah! These days it is slowly creeping up. But at the same time people also look at other aspects of the research like how difficult it is compared to others' work, how many people appreciate your work. It need not be number for example. You can read one of the papers in a leading journal quoted by one and all. That's enough. You don't have to publish 10 articles in the so-called low impact factor journals. But having numbers does not hurt. Churning out numbers is not a big thing in science now. You can always keep on churning out papers but ultimately people ask the question you know, what is it that you have done tell us in science in ten sentences. You have to come up with some thing."

The scientists further insisted on commenting how to be more productive, the scientist replied that, there is no end to the stock of knowledge. In his opinion societal issues must occupy the domains of normal science in a research paradigm with funding being diverted to such

research focusing on the immediate application to society. Consider the statement of the scientists:

> "Stock of knowledge? There is no end to it. I think the scientist should address societal issues because you know; those days are over, I am working for knowledge, knowledge has to increase, and knowledge is not there. Now more or less enough knowledge is there to do something more. I think societal issues should be addressed. Funding should be more directed towards that kind of things than doing something which is mundane, never heard of not seen so far kind of research you know. I am sorry those days are over. If it is something useful to the society, yes. It could be anything. I don't say this or that. I think that should be the way. I don't say immediate application to the society. That is true al over the world now. It is not there even in most developed countries. It should have an impact ultimately; it will certainly have an impact in the long run."

5.1.3 Number or Application

Some scientists from national laboratories insist upon the immediate relevance of the topic of research and its application to research rather than number of publications. He says that one has to create a knowledge base than increasing the number. He exclaims only a few publications are useful to common man's purpose and it is the latter that is very important. One of the scientists from a national R&D laboratory mentioned:

> On Mahatma Gandhi's samadhi, it is written, "Science without a human face is sin". Who ever does science, science should upgrade the society, march the country forward. In that case if science is to be translated into the society, the fruits of should reach the citizen.
>
> If some one says that he has published in this journal or that journal, ask him to explain how does it help the country. There he is caught. Just because I publish paper in JACS, JOC, NATURE it doesn't mean that I have mastered the subject. It has to reach the common man.
>
> The guys sitting at the helm of affairs should define what India, as a country requires and creates a knowledge base, knowledge pockets put application oriented research more important. When it come to numbers of papers, 1 of one million will be useful. 0.999 million are useless.

Some scientists said that due to one strategy of publishing, the Indianness of publication is questioned. He says that, in the Indian context, organic chemists refer to high Impact Factor Journals published in the west, and pursue similar research with minor variations. Thus the topics thus selected are specific to high priority areas in the west to serve their socio-economic needs.

Another scientist from a national laboratory says that the Indianness is lost, if the research paper addressed the socio-economic needs of the West. In other words, the research output is not going to be used in Indian in any way. A scientist from a national laboratory elaborates thus:

> "Best way of doing it is.... Look into journals of high impact factor, and carry out similar work in the same area or areas by changing some substituents or some conditions, carry out the work, write the papers and publish it in the same journal. This process goes generally well in most cases, because, the topics selected belong to *high priority* areas, selected, funded by advanced countries, depending upon their socio-economic needs. It does not reflect the demand and need of the country. No doubt if a big man sends, it will be immediately accepted, does not mean also depending on the relevant scientists contribution to the editorial board of the journal will accept or reject it.
>
> In India, unless you carry out the work related to high priority areas, the paper will not be accepted.
>
> Indian journals have low impact factor, nobody would like to carry out by preference or by choice any work that is suitable only to Indian journals,. All learned professors, renowned scientists of India, generally publish in journals of high impact factor, which by no means are Indian journals."

5.1.4 Productivity is Number! What about Teaching?

The perception of scientists about division of work in terms of teaching, research and administration seem to influence the definition of productivity. At IITs, and state universities, the scientists claimed that they have to spend more time in teaching. The IITs cater to the needs of variety of students enrolled in undergraduate, masters, students and Ph.D. programs apart from teaching the undergraduate engineering students. In the case of state universities major emphasis has been upon teaching. One of the scientists observed:

> "The productivity of a scientist is purely viewed or gauged in terms of the number of research publications that he has published or brought out already. That is the only way you can gauge or assess the productivity of a scientist. Of course, productivity means in research, I am talking about not in teaching, because research is the only one where productivity can be quantified. Okay, of course I have been teaching 4 hours a week or 5 hours a week for the last 20 years. But it is not a productivity-oriented job, it is a

creativity oriented job. I can come up with a lot of innovative methods in my teaching. But I have no way to prove it to you. Only my students know that I am a good teacher. I convey the material very well. But I have no written record of how good and how productive I am in teaching. In research I can view or judge a person purely by research publication."

Several scientists located in teaching institutions shared this view. These scientists said that teaching is a creative job, which shall not have a written record, of how productive a scientist has been as a teacher, hence, they emphasize upon number of publications as the productivity of a scientist. One of the scientists said:

"See there is no yardstick to measure"

5.1.5 Number with Variation

Some respondents say that in India there is a specific trend in which scientists conduct research on one particular topic and publish a paper. Later they make minimum changes in some aspect of the experiment, get results and publish a paper. Thus variation in the same experiment seems to be prevalent as a number churning game among organic chemists in India.

To quote a solid state structural chemist from the Deemed University,

"The number game is not a serious game; it is not that, you publish more papers in the same series…"You should not publish with minor variations."

5.2 Quality Aspect

In our study, we observed that, many scientists admit that, in the Indian context, it is still the number of publications that counts. Some of them suggested that the quality aspect of publication also must be considered. We asked the scientists about the quality aspect. Some scientists insisted upon indicators like citation index as an indicator of quality of the productivity of an organic chemist. Some scientists indicated that citation index may not be an accurate measure of quality of productivity as citation index is a function of citation practices..

There were few scientists who suggested considering the impact factor of journals, to determine productivity of a scientist. They too said that journal impact factor as an indicator of quality also has limitations.

A scientist from IIT suggests that the journals vary in terms of the impact factors. He quotes cases in which a scientist with large number of publications in low impact factor journal benefited with regard to the matters related to career, whereas those with relatively small number of publications in the high impact factor journals stood lower chances. This inequality in consideration of productivity, according to him could be dealt with multiplying the number of publications with respective impact factors. He stated:

> "yeah, I would suggest that, look into the value may be they can multiply each paper by the impact factor of the journal and then arrive at a number.
>
> Like NATURE, which has an impact factor lets say 9,10,11,12 or something and then you know one publication is equivalent to 12. Then you have one publication, which still meets the standards. Indian Journal of Chemistry by the way doesn't even have 1. It has 0.2,0.3,0.4…so even 12 publications come only to down to SIX. So you know you can probably think of it that way."

Some respondents advocate that the quality of the scientific work limited to number of publications and impact factor do have much significance for them. What is significant is, what impact could that work have on the scientific community. It is the quality indicator of productivity in science. Consider the following statements of one of the respondents:

> "That is why they have come up with these citation indices you know, I don't really believe that. There are many formulae like this. I don't think science is really such a thing. It can be quantified that way. It depends, e. g. even a Nobel laureate's winning work has never been quoted, that superconductor. It came just out of blue from somewhere. It had larger impact if not then impact factor is some thing you know something like X+Y= something. I don't believe this."

5.3. Factors of Productivity

In my discussions with the scientists I requested them to elaborate upon the factors that contribute to the productivity of the scientists.

Some respondents, especially from the state universities mentioned funding as a principal factor responsible for the productivity of scientists. They say that, on a comparative note, state universities, receive, nominal funding for research from the government and also allege that the scientists from elite organizations occupy positions in committees that allocate funding, tend to help scientists of elite institutions.

> "How do you explain the difference in productivity between State University and other institutions?
>
> It is like this. Central universities have more funding and they have more infrastructures. State universities have fewer funds
>
> They also get money from industry…but for Osmania University Government shall give only salaries. Increase resources from funding agencies…. We will do good research… and publish more papers".

During the visits to the laboratories of the scientists, I observed that, there was difference in the infrastructure both in quantity and quality terms between Deemed University, IITs, National Labs and Central University on one hand and State universities on the other hand. Respondents from the state universities said that, the other category of organizations has superior infrastructure compared to the state universities and that is why, they are able to perform better and are more productive. To quote an organic chemist from a state university:

> "Central universities, IITs and national laboratories they have super time facilities."

In this context, one may say that, from the respondents' point of view, technological infrastructure seems to be one of the factors responsible for productivity differentials among organic chemists in India.

Some respondents mentioned that choice of publication as an indicator of the quality aspect of the productivity of a scientist. Respondents mentioned that in the Indian context, generally that, scientists publish their work in journals that come from abroad, namely out side India, rather than in Indian journals. They claimed that the visibility of a scientist will be more if they publish in foreign journals than the Indian journals. Scientists claimed that the quality of papers published in Indian journals is low and refereeing in Indian journals is not rigorous. One of the scientists remarked:

> "See, take for example, TETRAHEDRON, synthetic communications., in their case, thorough refereeing takes place. Whereas Indian journals are like andhrajyoti and andhraprabha.(Local Telugu Weeklys)
>
> "Why don't Americans publish in Indian journal of chemistry? Why everybody in the world wants to publish in JACS? "

When asked to elaborate upon the reasons, one respondent said that the maintenance of a journal has become entrepreneurial activity in India. He says that, many journals exist though not recognized by knowledgeable academic circles. But these journals seem to accept almost every publication communicated to them. One of the scientists commented thus:

> "A paper rejected everywhere shall be taken by a journal of institution of chemists society from AGRA… JICS."

In the wake of number of publications being the established practice, such instances appear to be detrimental to Indian science and development in the long run.

5.3.1The Question of Patents

The scientists in our study included those who have patents. Patents are generally treated as different but equally important aspect of productivity. Patents are obtained for developing a product or a process by applying the existing scientific and engineering knowledge. There are problems in establishing parity between a research paper published in a refereed journal and a patent that has been sealed. One of the scientists remarked:

> "What do you say about PATENTS? Patents are hardly refereed, can you say that, you know I mean, there are no impact factors at all unless the company is interested in that. So how do you arrive at patents? The guy may not have publications, he may have patents also".

Some scientists in our study engaged in industrial research obtained patents. Scientists involved in industrial research in the national laboratories i.e, IICT and CLRI have to accept the mandate of these laboratories and focus on obtaining patents. The scientists in the labs expressed their concern. They stated that patents not being evaluated, the scientists who are employed in the applied divisions of organic chemistry at the loosing end to quote a scientist from a national laboratory:

> "Patents whether process or product are a donkey work…no time and money to attract high impact factor in the absence of high impact factor, the researcher ceases to be a scientist in the present Indian context."

5.4 Age and Productivity

I classified the scientists in terms of age groups and of five years each starting from the lowest of 25 years. We have categorized the productivity levels as the number of publications ranging from '0' to that of '50' and above, divided into 7 classes of 10 publications each (See table 5.4).

At the time of the study, the maximum number of scientists were in the age group between 40-44 years, followed by those in the age group 45-49 years. There is only one scientist below 29 years placed in a CSIR laboratory and there are 12 scientists above 50 years.

Table 5.4: Age and Productivity

	PRODUCTIVITY							
AGE	0	1-9	11-19	20-29	30-39	40-49	50 & > 50	TOTAL
25-29	-	1	-	-	-	-	-	1
34-39	-	3	2	-	-	-	1	6
40-44	-	-	3	8	7	2	-	20
45-49	3	3	-	2	-	4	2	14
50-54	-	1	1	4	1	-	2	9
55-59	-	1	-	1	-	1	-	3
TOTAL	3	9	6	15	8	7	5	53

I argue that age and productivity are related in such a way that productivity increases with age up to a point then reaches a plateau and slowly declines. The data does indicate this trend. Scientists between 40-49 appear to be at their peak of productivity (see Fig. 5.4)

This may also be the case with scientists in general in the Indian context. One may say that scientists in publicly funded institutions have to build their infrastructure and laboratory after getting their regular employment and start making contributions on a sustained basis. The initial phase may be slow, but picks up momentum after a couple of years or so. The younger scientists obviously spend more time initially to mobilize resources. Depending on the organization and its resource endowments, some young scientists mobilize resources and some can not. To quote a young scientist recruited in a state university,

Fig.5.4 Age and Productivity

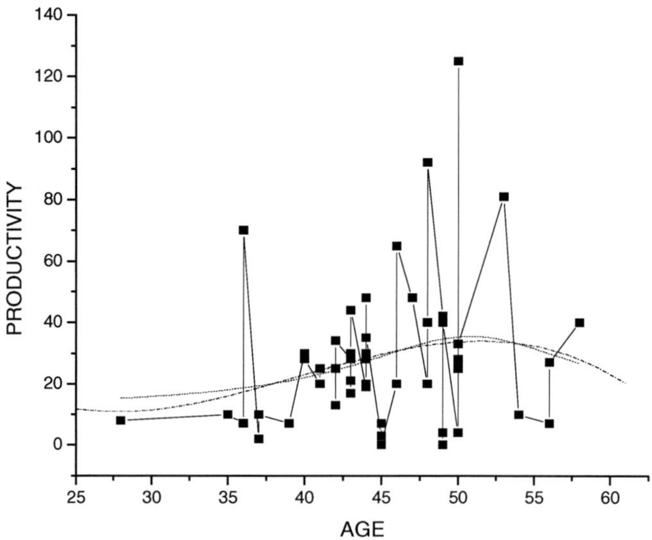

"I don't have lab, and without lab I can not think. In my doctoral institution, I almost spent my time in the laboratory itself. Here I have to look for laboratory, facilities, computer is not there. I cant check mail, What to do? I am going to one Dishnet near by. Now a days, on internet you get everything. The problem is you get everything on Internet. In holidays, I am planning to go to my doctoral institution, I can work in my previous lab."

5.5 Doctoral Affiliation and Productivity

In this work, I attempted to find out the influence of the institutional context of the doctoral work of a scientist on higher later productivity. As we mentioned in chapter 2, we considered number of publications during the five years (1996-2000) in refereed journals by the respondents as productivity (See Table 5.5)

Table. 5.5: Relationship between doctoral affiliation and productivity levels

DOCTORAL AFFILIATION	PRODUCTIVITY							
	0	0-9	11-19	20-29	30-39	40-49	50 and above	TOTAL
STATE UNIVERSITY	2	5	2	4	-	1	1	15
CENTRAL UNIVERSITY	-	-	-	8	1	1	1	11
IIT	-	-	-	-	1	-	-	1
CSIR LAB	1	-	-	-	1	-	2	4
DEEMED UNIVERSITY	-	2	1	2	3	2	-	10
FOREIGN UNIVERSITY	-	1	3	2	2	3	1	12
TOTAL	**3**	**9**	**6**	**16**	**8**	**7**	**4**	**53**

Fifteen respondents i.e. (28.30 %) have pursued their doctoral degree at a state university. Two of them never published. The majority among them *i. e.* 11 out of 15 has produced upto 30 articles. One respondent, a professor in the age group 50-54, with a higher average size of research group has over 40 publications to his credit during 1996-2000. The majority of the publications of the state universities appeared in national journals. Another professor who obtained doctoral degree from State University later worked as a postdoctoral fellow with a Noble Laureate in U. S. A in Organic Synthesis had joined the deemed university. He is the only one in this category who has 50 publications to his credit.

Twelve out of 53 respondents (22.64%) have obtained their doctoral degree from foreign universities. Four among them are highly productive having published more than 40 papers during 1996-2000. Another 4 of them published 20-40 publications and 3 of them had publications upto 20. Only one scientist who studied abroad, aged 39 and currently employed in a national laboratory has published less than 10 papers during the five-year period i.e., 1996-2000.

Eleven respondents out of 53 (20.75%) who obtained doctoral degree from central universities. A majority (8) of them have published between 20-29 followed by 3 who published above 30 articles with one each in 30-39, 40-49 and 50 and above category.

Four respondents, (7.54%) out of 53 had affiliation with CSIR labs during their doctoral program, two of them being highly productive with over 50 papers. The duo-a senior scientist and a junior scientist-located in the CSIR lab have been collaborating for quite some time as the junior scientist's doctoral work was supervised by the senior scientist. The junior scientist continues to collaborate with his mentor even after obtaining his.

Only one among the respondents pursued his doctoral degree at an IIT and was inducted into deemed university. He belongs to the high productivity category with approximately 30 publications to his credit. Five respondents who graduated from the deemed university published between 20 and 50 articles. Three scientists who graduated from deemed university and later recruited by the state universities and national labs published less than 20 during the five-year period.

5.6 Post Doctoral Training and Productivity

In this section, I examine whether the difference in terms of postdoctoral training outside and within India had any association with the productivity differentials among organic chemists.

Postdoctoral training seems to be a transient stage in which scientists tend to have an opportunity to perform multifaceted roles and experience a kind of transformation from the status of an individual guided by a mentor to that of an independent researcher. Postdoctoral training facilitates an individual trained in a research area or a paradigm to shift over to or experience different area of research and methodology. A professor from IIT said:

> "Postdoctoral study is very important because it allows you to **cross-fertilize yourself**. You get to know, spend some 5 years in one group and move into another group and **practice different methodology**. You get to know 2 different ways of approaching a problem. PDF work is very important because later when you go start to work assuming that the working place will be different from these two. Then you have to experience of working at 2 different places, which you can put together and start looking at things in a different manner."

Thus it seems postdoctoral training enables a researcher to think independently and expand the limits of knowledge and gain expertise of a scientist before he joins a regular career.

Doctoral training may require a scientist to work in one area of research. This is indoctrinated while being enrolled into the research group. They work on a research problem in one subject for longer duration. By the time the doctoral training is complete the discipline itself would have undergone a revolutionary change and also could have given rise to a new sub-field. Thus, postdoctoral training enables the scientists to plug gaps in one's knowledge in a specialty.

Training in scientific research is immensely related to the knowledge of using existing gadgets and fabricating experimental apparatus etc., Technology of research- mastering experimental techniques and methods and technology associated with research-technological infrastructure and experimental techniques change due to continuous innovations in techniques. Post doctoral training may also be significant from the point of view of the quality of infra structure. Scientific research requires different kinds of infrastructure. The infra structure ranges from a simple to complex form. It also varies in terms of quantity, quality and efficiency. Different scientific organizations contain varied infra structural facilities of different levels of sophistication. A scientist who worked in a laboratory with one kind of technology moves to another as the latter is endowed with latest technology like sophisticated research equipment. Thus by moving to another laboratory endowed with better technology, as a postdoctoral fellow, a young scientist makes up the technology lag he/she experienced at the institution where he/she carried out his/her doctoral work. Scientists in our study emphasized significance of better quality infrastructure. Some of them mentioned that it is the quality of infra structure, that attracts scientists for postdoctoral training.

In the study it is evident that, majority of the respondents received postdoctoral training abroad, especially universities, in the west, like United States of America. They claim that the laboratories abroad, especially in the west, are equipped with better infrastructure and technology related to research. Thus many scientists from Indian scientific organizations prefer to undergo postdoctoral training abroad to fill the lag in technology and pursue research with better technology.

> "Essential to understand how they organize themselves in doing science. Writing proposal.
>
> Coming up with **creativity** Opportunity to work in a frontier field Exposure to **latest technology**. Mixing with various kinds of people with science as a common interest"
>
> "If you want to improve yourself you have to see the world."

During the doctoral years, the students are paid a fellowship for maintenance exclusively. The post doctoral fellows are offered salaries, in some cases, more than even a scientist obtains after recruitment in the Indian academic institutions. The quantity of financial assistance depends on the budgetary support extended to the organization or laboratory or project. Respondents from the State University state that their students go abroad because they get good salaries as postdoctoral fellows. Some respondents mentioned that postdoctoral training represents a transient phase in the professional career of a scientist after the doctoral years.

Doctoral years represent a phase, in which a student is dependent on and learns many aspects from the supervisor. The doctoral student is taught and is expected do his doctoral research work. Thus, doctoral period represents a phase of dependence. Postdoctoral training appears to be a transition phase wherein a scientist works in a laboratory with a senior scientist. He, according to some respondents from the state universities, advises the doctoral students in the laboratory, oversees the research work of the other students and also contributes ideas to the ongoing research work in the laboratory. Thus postdoctoral training in science seems to be a transient phase in which a young academic doctor metamorphasizes from the role of an dependent learning student researcher to that of an independent, advising, supervising researcher. To quote a Professor, from a state university,

> in doctoral training you are supposed to take courses and do some original piece of work. So, of course he is rather guided and taught by guide in every aspect. He is supposed to be undergoing you know, his guide is supposed to be teaching. But you know at postdoctoral training, the guides of the postdoctoral work are not expected to supervise his work on day to day basis. He has already been trained over a period of four and five years. So now it is an additional training for him to work independently. Do you understand what I mean, "work independently" and help other students. A postdoctoral mentor may have some other Ph. D students. It gives him an opportunity to supervise the research work of other junior research students. Now he him self is senior Ph. D okay, so this gives him to take care of a research group. Take care of undergraduate postdoctoral and coming doctoral students in the initial stages to teach them, to guide them and if possible if he is a very creative Ph. D, he can even contribute his own idea. This he would have done even in his student days, but now he has more opportunity, so postdoctoral training is a must rather for anybody."

5.6.1 Post Doctoral Training abroad and Productivity

I observed that out of 53 respondents, 33 have been postdoctoral fellows outside India. Among those who have been postdoctoral fellows outside India, 24 scientists were postdoctoral fellowships for duration between 1 and 4 years (See Table.5.7.1). Among the scientists with postdoctoral duration between 1 and 2 years, two scientists have up to a maximum number of 19 publications. The other 11 scientists published between 20 and 50 publications. In the category of scientists with postdoctoral duration of 2 to 4 years of postdoctoral training, one scientist had no publication at all and one had less than 9 publications. The other scientists in the category published 20 to 50 papers.

Table.5.6.1: Relation between duration of Postdoctoral Training abroad and productivity levels research publications (1996-2000)

PDFADU	0	1-9	10-19	20-29	30-39	40-49	50 AND >	Total
0	2	5	3	5	3	1	1	20
>1&<2	-	1	1	3	4	2	2	13
>2&<4	1	1	-	4	1	3	1	11
4 & above	-	2	2	3	-	1	1	9
Total	**3**	**9**	**6**	**15**	**8**	**7**	**5**	**53**

PDFADU: Duration of Postdoctoral Training Abroad

It is clear that the data tabulated above is corroborated by the perceptions expressed by individual scientists during the course of the interviews that postdoctoral training abroad shapes a person to be a more productive researcher. Some scientists during the interviews elaborated upon the advantages of postdoctoral training abroad.

At the level of the organic chemists in the study, choice to undergo post- doctoral work in a foreign country is considered highly desirable. Thirty-three (62.26%) respondents pursued post- doctoral research as post doctoral fellows, research associates in various universities and research institutes in foreign countries. Scientists reflect this across different organizations under study. The duration of post-doctoral training ranged between, less than one year to six years. The state universities (9 respondents) and national laboratories (9respondents) and 2 scientists from the central universities did not have postdoctoral fellowships.

The respondents were asked to indicate as to why majority of the Indian organic chemists preferred to go abroad for post-doctoral research and training. The reasons that emerged out of the responses are given below.

- Cross- Fertilization of ideas
- Theoretical and Experimental Science
- Quality of Research
- Technology Lag
- Quality of Research

- Research experience in different countries
- Related to career and recruitment

According to some respondents, postdoctoral training abroad creates an opportunity to interact with individuals with different kinds of ideas. Such a cross-fertilization of ideas they claim contributes to the progress of science. To quote a scientist from an IIT,

> In both cases, I see if they go abroad, especially since we are talking about cross-fertilization. . I.e.... we are talking about as diverse background as possible, going abroad is slightly more beneficial in terms of you know getting more view points, an so on…

Some Organic chemists distinguish between theoretical and experimental research postdoctoral training. They feel that postdoctoral training doesn't make a difference for those scientists involved in theoretical sub disciplines. Says an organic chemist from the central university,

> "Depending on the discipline of course, it should be better also yeah for example I don't see anything non-experimental science abroad is better than what it is in India, I don't think so, mathematics for example, if people have to go abroad and learn anything, of course, experimental sciences are there. Physics, some equipment thing… there is a difference in the training pattern that doesn't mean that Indian training is bad or abroad is better. I only perceive that it is different other than that I don't think abroad training specifically helps somebody to do better science in India. In fact, the other way round may be true. Indian training and going abroad to see what is there then come back is a better idea. Then you can compare both of them"

They claim that in experimental science, postdoctoral training abroad is advantageous to scientists, as the laboratories in India have limited facilities and infrastructure.

Some respondents in the study cited the advantages of being postdoctoral fellows abroad with well-known scientists who are established authorities in different research fields. They said that, by being a post doctoral trainee with the 'big man' in the field, a young doctorate will have advantages like high quality peer interaction, exposure to the best of the sources of knowledge and technology and good 'contacts'. He would be part of the invisible college, which will have implications for research in future.

Scientists spoke about the importance of social conditions especially the quality of facilities required for every day life and their impact on the progress in research work. Scientists claim that in India, they suffer for want of time, as time, the essential element in research and progress of science is wasted due to frequent power failures, and other things like less organized service sector. They claim that living conditions in countries outside India, especially the West, are better. Social life is better organized and conducive for research.

5.6.2 Post Doctoral Training in India and Productivity

In our study we found that 79.25% of the respondents have never pursued their postdoctoral research in India. Of the remaining 11 scientists, 8 have done their postdoctoral study in India for duration of one year each. They have up to a maximum of 39 publications. One women scientist from State University pursued her postdoctoral study for five years and has less than 10 publications to her credit. Another scientist from the Central University had pursued his postdoctoral study in India for 2 years before he went abroad for the same for another duration of 2 years. (See Table.5.6.2)

Table.5.6.2: Duration of Postdoctoral training in India and productivity

PDFID	PRODUCTIVITY							
	0	1-9	10-19	20-29	30-39	40-49	50 AND >	Total
0	2	6	5	12	5	7	5	42
1	1	2	-	3	2	-	-	8
2	-	-	-	-	1	-	-	1
3	-	-	1	-	-	-	-	1
5	-	1	-	-	-	-	-	1
Total	**3**	**9**	**6**	**15**	**8**	**7**	**5**	**53**

PDFID: Duration of postdoctoral training in India

He has to his credit, above 30 publications. Another scientist from a national laboratory with a postdoctoral experience of three years belongs to the category of those who published between 10-19 papers.

Thus it appears that scientists in the study preferred to train themselves abroad after the completion of their doctoral degree rather than in India. Those who received postdoctoral training in the Indian context seem to show low productivity levels in terms of number of publications.

I examined whether scientific organizations in India recognized the importance of postdoctoral training and consider the same when they recruit scientists. We observed form the interviews that, post-doctoral training is being considered in every scientific organization, except among the state universities. Scientists from the IITs and the central university agreed upon the fact that these organizations consider a minimum duration of three years of postdoctoral training required as a norm for recruitment at entry level. To quote a scientist from an IIT,

> "We do. In IIT we can not take any one as an Assistant Professor unless you have a PDF experience. It is given on the you know requirements itself. We need 3 years of PDF experience after your. to become an Assist. Professor, If you have just. you can become a lecturer, which is not, a regular job here. It is only a contract job. So to become a faculty member here you need PDF experience."

I also tried to examine the social aspect of influence of the postdoctoral mentor, when a scientist applies for a job. Scientist from the deemed university said that even among the post-doctoral fellows, those who were postdoctoral fellows with Nobel laureates are being considered more seriously at the time of recruitment in India and this trend must change. To quote him:

> "It seems to matter until recently students of Nobel laureates used to get positions. This should change. We should have peopled who want to work in India. In our country, infrastructure is of low quality, bookish knowledge, and no exposure to modern day facilities. In fact it is openness that matters."

On the other hand, there were also the scientists who said that postdoctoral training and the mentor may not really matter for promotion. They felt that for promotions, the person with whom one pursued PDF training does not come into picture but it is an internal process involving scientists within India and the concept of 'God Father' still works if one joins the same department where his mentor is still an active member. Says a scientist from the Central university,

> "I don't think so. It depends on where he joins. For example if you join the same place where you do your Ph.D. . yes, may be because there is the same supervisor he has a godfather. Otherwise I don't see any difference in studying abroad or studying here."

Thus, it appears that, while contacts and networking with the doctoral organizations help in promotion, association with highly rewarded scientists like Nobel laureates seem to work for an entry-level employment.

In our study we found that some state universities did not have recruitment for at least 10 years. Scientists, from the state universities said that, absence of recruitment is one of the factors, which led to the flight of doctorates from the state universities to foreign countries for postdoctoral training. A Professor from a state university observed:

> "There is no use in doing PDF in India. After., If one gets a job, nobody applies for a post-doc. Doing PDF abroad has become a vanity... craze... People go for PDF just like that. Even if the money is less, some go there to get a job with H1-Visa. Because there are better opportunities. Less than one percent of the people go to famous people. Who would not want to go and work with a big man, expose themselves to frontier areas of chemistry?
>
> There are no jobs for Ph.D. here. So it is an outlet. For the last 10 years, There has been no recruitment. . Ph.D. Output is very high. Where will they go? We lost best students. Has there been recruitment, Quality of the department would have increased.

They say that those who go abroad for postdoctoral training look for a career in the industries abroad or they come back to India to join industries as scientists.

> "Abroad, they have good resources, the scholar is given a salary, i.e., more than he requires. There is good infrastructure. Whereas, we are a poor country. Getting a permanent job is very difficult. There is no comparison at all. If anybody does a PDF in India he gets good knowledge but getting a job is difficult."

Thus post-doctoral training abroad especially from the point of view of the state universities provides a loop line for a career in Industries both in India and abroad, in the light of the absence of recruitment into academic positions.

5.7 Duration of Employment and Productivity

The context of the organization, where one is employed as a researcher provides resources:- physical, human and cultural. Physical resources relate to infrastructure like space, power, water etc., Human resources refer to research students at different levels who work with scientists. Cultural resources include books, journals that provide ideas, and interaction with colleagues that help in research. The organization also provides a framework of norms that facilitate interaction with students, colleagues and bureaucracy in the pursuit of teaching and research. However, sometimes the norms may also inhibit research due to their inflexibilities. Scientists have to negotiate with enforcers of norms to evolve interpretations of norms to carry out their work. The duration of association with an organization that provides the above mentioned seems to exercise some influence on the performance of the scientists' productivity level. Hence, we intended to find out to what extent duration of association of the scientist at an organization has an impact on the productivity. (See Table.5.7.)

Table 5.7: Duration of Employment and Productivity Levels

DURATION	PRODCUTIVITY							
	0	0-9	10-19	20-29	30-39	40-49	50 and above	TOTAL
0-5	-	4	1	-	-	-	-	5
6-10	-	1	1	2	-	-	1	5
11-14	-	-	3	7	6	2	-	18
15-19	2	3	-	3	2	3	2	15
20-24	1	-	1	3	-	1	1	7
25-29	-	1	-	-	-	1	1	3
TOTAL	**3**	**9**	**6**	**15**	**8**	**7**	**5**	**53**

The duration of affiliation seems to have two dimensions: one, the number of years; and the other, the qualitative one. One may categorize the qualitative dimension into three phases. The first being, the **naturalizing phase**- that involves socialization and assimilation into the respective social system**,** the second, **Performance phase**- that involves appropriate use of human, organizational and knowledge resources to produce knowledge, and the third, late **phase** wherein either scientists disassociate themselves from the research activity or allocate relatively less time as they are drawn into administrative activities of the organizations.

For purpose of analysis, the duration of association is divided into three class intervals: up to 10 years, 11-19 years and 20 years and above. Ten respondents, who worked at the current organizations for less than 10 years, and have to their credit, 30 publications. The duration that ranges between 11 and 19 years seems to be the most productive period of scientists characterized by showing peak performance. This period may be divided in to two sub-phases. Thirty-three out of 53 belong to this category. Out of 33, 18 respondents worked for 11 to 14 years followed by 15 scientists in the second sub-phase that worked for 15 to 19 years. While those in the first phase published between 10 and 50 publications, some in the second sub- phase, have published more than 50 publications.

Ten respondents spent over 20 years. Among these, two scientists from state universities have published upto 20 articles, 3 scientists published around 30 and the rest 4 had above 40 publications to their credit. Of the 4, two are from central universities and one each from CSIR lab and the deemed university.

Thus, it appears, in case of IITs, Central universities, CSIR laboratories and deemed universities, among the other factors, duration of stay is associated positively with productivity of the organic chemists.

Fig. 5.7 Duration and Productivity Curve

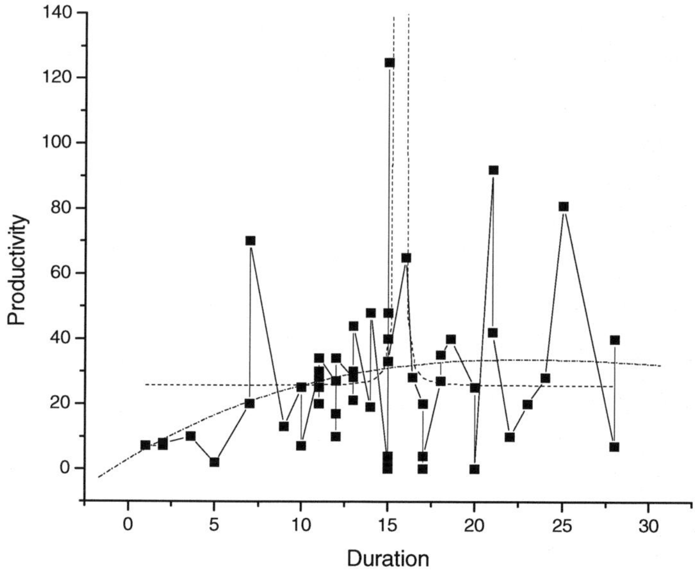

5.8 Similarity of Research Area and Productivity

I mentioned earlier that the stage of cognitive development of a research field influences productivity. If the field is new it provides lot of opportunities for research, both in terms of quantity and quality. Research specialties show different phases of growth. New theoretical insights or a new paradigm opens up possibilities of new research agenda. Initially, research grows rapidly, later reaches a plateau and then slowly declines. The opportunities for research in the field follow this trajectory. In this study, we asked the scientists to indicate if they shifted to a new area of research different from the area of their doctoral research. They were asked to indicate on an ordinal scale, the extent of similarity in terms of a similar to very large extent; to a large extent; to small extent and to a very small extent. The idea here is that, if any scientist shifted to a different area, that area may have been a new area with a lot of opportunities for productivity in terms of quantity and quality. The extent of shift in the area of interest is then related to their productivity and cognitive development of a research specialty. (See Table.5.8)

Table 5.8: Similarity of Doctoral Research Area and Present Area of Research

SIMILARITY OF RESEARCH AREA	PRODUCTIVITY							
	0	0-9	11-19	20-29	30-39	40-49	50 and above	TOTAL
TO A VERY SMALL EXTENT	-	1	4	2	-	3	3	13
TO A SMALL EXTENT	1	1	1	2	-	1	-	6
TO A LARGE EXTENT	1	1	1	7	-	1	-	11
TO A VERY LARGE EXTENT	1	6	-	4	8	2	2	23
TOTAL	3	9	6	15	8	7	5	53

Twenty-three respondents reported that their current research area was similar to the area, at the time of doctoral work to a very large extent, of which 6 scientists were from State University with productivity level between 16 to 20 publications. One respondent aged 49, from the applied division of a national laboratory has no publications.

Eleven respondents reported similarity to a large extent, 7 of them falling in the category of 20-29 publications. One published above 40 and 2 below 20. One respondent aged 49, from a national laboratory has no publication despite continuing to a large extent in the similar area of research as to that of doctoral age, despite a professional age of 17 years.

In the case of 6 scientists, similarity of research is to a small extent. One of them, a scientist aged 48 from central university published above 40 and 4 between 10-20. One scientist aged 48 from a national laboratory had never published. Thirteen respondents reported similarity to very small extent, of which 6 rated above 50 productivity. Another 6 have 10-30 publications. One respondent aged 58, from a state university has below nine publications.

It appears that there are two trends, one, a very high degree of similarity which appears to be conducive for higher productivity providing a large scientific pool for exchange of ideas and information, opportunity for collaborative research within and outside the country. The second trend with a very small degree of similarity wherein the scientific pool is less in size and may be resulting in:

1. Collaborating with scientists from other disciplines, or
2. Taking up different research area
3. As per the organizational mandate after being inducted into the particular organization.

Nevertheless this category of scientists continued to be consistently productive due to other factors that would be identified in further analysis. As mentioned above, shifts in research interests may be due to opening up of a new research field with greater opportunities for research.

5.9 Workgroup size and productivity

We categorized the average size of the research group of the scientists into 7 classes with an interval of 5 each. We categorized the productivity of the scientists into 6 classes with 10 publications in each category in ascending order (see table 5.9)

A majority of the scientists with an average size of the research group between 1-5 published between 0-50 publications. This trend is followed by the scientists with an average size of the research group 6-10 published between 10 and 50 and above articles. A small group of scientists with an average size of the research group 11-15 published between more than 30 articles in refereed journals.

Table.5.9: Average size of the research group and productivity

Group Size	Productivity(No of publications in refereed journals between 1996-2000)						
	0-9	10-19	20-29	30-39	40-49	50 and above	Total
0-5	9	4	8	2	3	1	27
6-10	-	1	6	5	5	4	21
11-15	-	-	-	2	-	1	3
16-20	-	-	-	-	-	-	-
21-26	-	-	-	-	-	-	-
26-30	-	-	-	-	1	-	1
30-34	-	1	-	-	-	-	1
Total	**9**	**6**	**14**	**9**	**9**	**6**	**53**

Fig.5.9 Average size of the research group and productivity

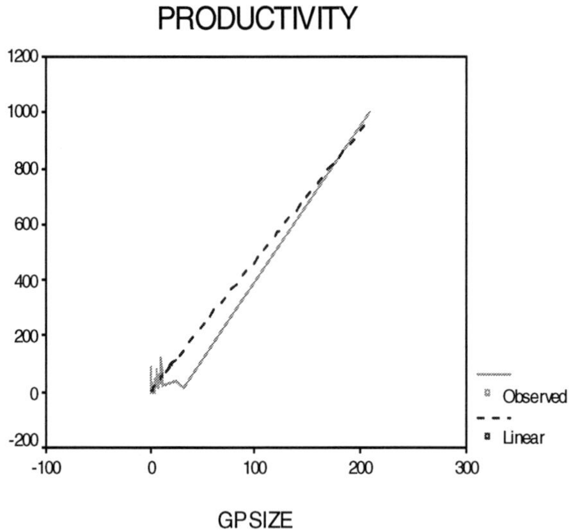

One scientist with in the class of 26-30 size of the research group published above 40 articles in refereed journals. One scientist from the state university with maximum number of students in the research group with above 30 members has published less than 19 articles. It appears that (See Fig 6.5) with the increase in the size of the research group of the scientists, there is an increase in the productivity levels of the former.

5.10 Summary of Findings

I explored the perceptions of the scientists about what constitutes productivity in the Indian context. We observed that though the scientists in the Indian context consider quantity and quality aspects of research publications as important dimensions of productivity, the existing social structure in Indian science seems to emphasize upon number of publications.

Respondents felt that productivity is difficult to define from the point of view of research area being relatively small or new. Such a situation it self is a limitation as it offers a few knowledge peers comprising a small scientific pool. In such a case there is less scope to publish and the number aspect may not help to define productivity. Some scientists who seem to observe that the quantitative aspect is incomplete in determining the productivity level of scientists say that, to balance, the number of publications, must be multiplied by respective journal impact factors and added to make a score or citation indices must be taken into consideration.(Policy implication). Some others also suggested that the impact of the research work on the scientific community must be considered. Scientists seem to emphasize upon the impact factor of the journal and citation index to assess the quality of productivity of the scientists. However, they mentioned that these are just emerging in Indian science.

Scientists employed in the national laboratories that are mandated work out applied research, can not publish to the extent as scientists in a university. But file patents on the basis of their applied research. In the Indian context it is observed that patents are hardly refereed. Hence, they insist that scientific productivity must be measured from the point of view of immediate application or relevance of research work to common man.

Scientists from IITs insist upon considering the number of teaching hours while calculation productivity as they the organizational mandate requires them to spend relatively more time on teaching undergraduate students. In contrast scientists from the state universities say that teaching is a creative job with no reliable measure of how productive a teacher is. In such a situation, they say that in India it is still the number of publications that is considered as productivity.

Interviews and discussions with scientists on factors responsible for productivity of the scientists revealed the following:

According to some scientists, the sizes of the scientific pool in a given research field available for interaction influences the productivity level.

The higher the number of practitioners in a research area the higher are the chances to be more productive.

Scientists from state universities cite, that, the quantum of funding as a factor that influence productivity. They said that scientists from elite organizations occupy positions in the review committees that allocate funding. They favor individuals and organizations that are already well endowed neglecting scientists from less resourceful organizations such as state universities. Thus inequality in scientific productivity is related to differential allocation of resources, favorably awarding funds to the already reputed individuals and institutions. One can see the 'the Mathew Effect' in operation.

Productivity differentials are accounted for by the quality infrastructure. High quality infrastructure positively influences productivity levels. This is corroborated by my observations during the visits to the labs.

The number of publications as the quantitative aspect of the productivity suffers due to the emergence of low quality local journals that accommodate almost every research paper communicated to them.

The productivity level of scientists varies with age. Except for one scientist, scientists below 40 years belong to low productivity category. In the age group of 50-59, 50% of the scientists belong to the middle level of productivity and 25% each belong to the low and high levels of productivity. Our study shows that productivity of scientists slowly rises and reaches its peak between 41-19 years and thereafter reaches a plateau and then slowly declines.

For the sake of convenient analysis we classified productivity levels into three broad categories. They are low productivity category, middle level category and high level productivity. Scientists who received doctoral training at the state universities fall in the low productivity category. Scientists who received doctoral training in the central universities fall in the middle and high productivity category. The only respondent in the study who graduated from an IIT belongs to high productivity category. Scientists who graduated from the deemed university fall in all categories of productivity. Scientists who graduated from foreign universities fall in the category of middle and high productivity.

The optimal duration of postdoctoral training seems to be around 2.5 years. Those who receive postdoctoral training below one year and above 4 years do not seem to be more productive. Scientists who received postdoctoral training outside India belong to the middle and high levels of productivity.

Of the respondents who received post doctoral training in India 80% received it for one year and only 20% received postdoctoral training for more than one year. Only one scientist who received postdoctoral training for two years was in the middle level of productivity and the rest in the low level of productivity.

The duration of association of a scientist with current organization employment seems to influence productivity levels. Scientists who are associated with current organization of employment for less than ten years fall in the low productivity category(21.2%)

The optimum duration of association with an organization seems to be 11 to 19 years wherein we observed that scientists reach starting from middle level through high and to the highest level of productivity. Those who are associated for more than 20 years with organizations, 60% of the respondents fall in low productivity category and the rest 40% in high productivity category.

CHAPTER SIX

COMMUNICATION, COLLABORATION AND PRODUCTIVITY

Introduction

This chapter is divided into two sections. In the first section I shall discuss the pattern of communication among the organic chemists in the study, science communication, conferences in science, with their utility to scientists and students and publication behavior.

In today's world computer-mediated communication (CMC) has pervasive influence. Scientific communication is also influenced by CMC. We shall examine the role of communication technology in scientific research. I describe the shift to electronic mail based communication, namely the computer mediated communication technology, adaptability to the same by scientists from different organizations and age groups, impact of electronic communication on communication, both formal and informal. We describe the relative contribution of electronic scientific communication among other factors to the productivity levels among the scientists.

In the next section, I describe the pattern of collaboration among Indian organic chemists. We address issues, such as what is the basis of collaboration in the Indian context, why scientists collaborate, whom do they collaborate more with and why? What are the relative advantages and disadvantages of collaborative research? I specifically examine the relationship between collaboration and Productivity.

SECTION-I

6.1 Communication in Science

Communication in science as in any other human activity is vital. It starts at the level of an idea. The ideas are precursors to knowledge. We explored questions related to modes of acquisition and transmission of ideas. We attempted to find out the channels of communication of ideas, types of communication deployed by the scientists in the study. We categorized the responses into human and non-human factors. We examined the role of communication technology in translating the ideas into knowledge.

Among the human factors, we have identified two different kinds of factors, individual and group. Scientists responded that among the individual factors, serendipity is the one that leads to new ideas.

> "What is the source of ideas to a scientist?
> Serendipity… flash…."

Some scientists from national laboratories say that informal discussions with colleagues and some spontaneous ideas contribute to the development of the ideas that initiate the production of knowledge.

> What is the source of ideas to a scientist?
> Spontaneous, informal, colleagues, library and journals.

Some respondents emphasize upon the fine-tuning of existing research. They say that more probes into the existing research problem shall result in the generation of new ideas in scientific research

- Continuation of existing research, finer aspects, fine tuning of your research
- Consolidate old ideas.

6.1.1 Lab as a site of dissemination

At the deemed university in one research group there is an organized mechanism in terms of group discussion. All the students in the research group meet for once in a week, i.e., every Saturday and discuss the literature they reviewed. In addition, the mentor also holds discussion with individual students. To quote a scientist from a deemed university:

"In our lab our students review the literature, and every weekend we have a group discussion. Saturday we sit and discuss.

Individual discussions with students also take place…"

6.1.2 Journals

The majority of the respondents say that journals are the sources of ideas and they suggested going to a library and reviewing the journals, as the best form of generating ideas related to research. To quote a scientist from an IIT,

> **Q)** Having access to Internet, does it help a scientist to get new ideas? Or does it help the scientist to get the information about the in things in the field?
>
> **A)** "I don't know in science, there are two things,
> One is going to the library, I am talking about gathering information, one has to go to the library, read journals, be it 2000 or be it 1950 you have to read the papers, you have to read everything, to get the information, and the second - is going to conferences, of high quality where people working in that field, come there discuss, bring new results, and so on. These are two different things. "

6.2 Conferences are Communication Events

In scientific research, there are different ways of publicizing the outcome of research work. They are: publishing the research in a refereed journal and presenting the findings in conferences, the institutionalized arrangements to recognize the knowledge produced . Conferences invite contributions both from scientists and students. Conferences are occasions to exchange ideas. We asked the respondents what purposes do conferences serve in science, both for scientists and the doctoral students.

Conferences are formal occasions during which scientists communicate their research to their peers. Such communication helps establishing priorities and also getting evaluation and feedback on one's worth from the peer group. In addition to this formal communication during a conference, scientists have an occasion to meet face to face informally and exchange ideas. Informal communication is an essential aspect of science, to help identifying potential collaborators on the basis of complementary strengths- theoretical, methodological and technological. To quote a scientist:

> "Conference is a medium by means of which people exchange ideas…"
> I went to Trivendrum for a conference titled "Recent trends in photochemical sciences". Participants have come from all over the world, which included 50 foreigners and may be another 50 Indians. At conferences people can exchange ideas during dinner and lunch through informal discussions. You are free to discuss off the dias. Other person also receives in a proper way.

Scientists constantly extend horizons of research resulting in the emergence of new areas of research. The new research specialties are promoted by making them as the conference themes. Scientists, who work in other areas of research, by participating in such conferences, get an opportunity to have exposure to these frontier areas, and some scientists may eventually shift to a new area of research. A scientist echoed this:

> "By attending a conference a scientist can get into new fields and collecting new ideas is another advantage. Collecting knowledge in the frontier of knowledge in the new field increases."

As mentioned above conferences, provide opportunities for identifying potential collaborators on the basis of complementary strengths. For example, scientists whose labs are endowed with new technologies, decide upon to meet others who have the technology and other resources. Those who do not have the appropriate technology have an opportunity to meet the 'haves' and forward a collaborative proposal. Thus, we could say that conferences are occasions to initiate collaborative research. Says a scientist from a national laboratory:

> "Collaboration can be initiated.
> Suppose I don't have the instrument, which he has with him, I impress upon him, he can do the work for me. That is how you can go for a collaborative project."

In science, there exists a hierarchy among its proponents in terms of knowledge, competence, rank, and power (academic authority). Some senior scientists in the discipline occupy advisory positions in the national and international funding agencies, and government bodies.

For a young scientist conferences may provide opportunities to establish rapport with senior scientists who may suggest potential funding agencies and the thrust areas that can attract fund. Young and motivated scientists may take cue from a senior scientist and take necessary follow-up action to mobilize funds.

"You can catch big guys for funding"

6.2.1 Conferences and Research Networks

Scientists are located in different organizations. By attending a conference, they come in touch with others' in the research field and gain acquaintance regarding their expertise. As mentioned above, complementary strengths form the basis of future communication and collaboration networks. This contagion process as Diana Crane(1979:70) describes is the first step towards the emergence of a network in a discipline. These contacts, as the respondents say, are essential, as the research endeavor is a collective pursuit. Conferences also provide an opportunity or a forum to facilitate social interaction between the leaders, on the one hand, younger scientists as neo-phytes in science on the other.

> "One can develop relations with various kinds of people from different countries.
> Helps to develop contacts with other scientific community."

To quote scientist from state university,

> "For scientists they are occasions for social interaction."

In our study, we found that, conferences serve different purposes, for young doctoral students. Doctoral students as members of research groups are neo-phytes who are in the process of getting socialized into the culture of the profession. For them conferences provide an occasion for:

1. Professional anticipatory socialization.
2. Opportunity to enter the scientific network or initiation into the networking process.
3. Institutionalized opportunity for exchange of information and ideas.

6.2.2 Conferences and Priority

In science, pursuit of production of knowledge is a continuous process. The process results in innovations, discovery and many more facts observed. Traditionally, scientists as per the Mertonian paradigm make claims over priority by publishing their work in refereed scientific journals. The whole process of sending the manuscripts, refereeing and publishing according to some scientists is a long, cyclical process according to some respondents. They say instead, by way of presenting the results in conferences, scientists could make early claim, without losing much time in the process of making knowledge public. To quote,

> "A student gets an opportunity to present before an august gathering. If he could convince, he has better opportunities., It is good training. In fact, doing is different, explaining it is different, which helps him to become a good teacher. He must explain why he used the concepts and how he used the methodology. "

6.3 Modes of Communication: Transition to Electronic Mode

Technology has been an integral part of society being an essential component in many social institutions. The advent of information technology seems to have significant impact on scientific research. Communication in science namely, exchange of information among peers, correspondence between scientists editors of journals, referees, with funding agencies, has been an integral aspect of science as a social system. For ages, scientists resorted to different media of communication like surface mail, telephone, and fax in science communication. The advent of information technology, triggered by the invention of computers, networking and electronic mail, seems to have a significant impact on the social structure of science. Electronic mail seems to emerge as a dominant mode of communication.

Interviews revealed that organic chemists in our study use different media at different stages for scientific communication. Scientists say that, surface mail is used. Many of the 53 respondents completely take recourse to electronic mode of communication at every stage and narrated their experience how e-mail facility helped transmission related to collaborative work sent for publication in a very short duration. They say subsequent to the acceptance to be published, and other works like changes in manuscripts and all are being sent through non-electronic traditional modes of communication. One of the scientists mentioned:

> "No. For communicating Manuscripts, I think Surface mail is still used most. Of course we have submitted some manuscripts online. There are some journals, which have these facilities to submit manuscripts online. We have in fact done something and we have received, I know, the acknowledgement, the number and we are handling this and so on. So you know, that would be a very ideal thing because that leads to what you call a paperless office. We haven't seen anything and in fact we have done it for few journals also. Like our journals in TETRAHEDRAN, which was, I can

tell you a minor incident. You know, like I was in Hamburg, two years back. My student, who just finished, a post-doc in Netherlands. So, one week end he came to me with the manuscript and everything, we typed the manuscript, we did everything at the Hamburg lab on the computer and then we sent it through e-mail straight from there to the journals editorial office. That Monday I got the acknowledgements. And two weeks from that, when I was in Berlin, when I checked my e-mail, I could see the reply from the editor saying that it was accepted with minor revisions. So we incorporated these revisions in Berlin lab and then we sent it. I came back to Hamburg, I got an e-mail saying that the paper got accepted. So here, in two weeks time, we finished everything without exchanging a piece of paper. I did not sign anywhere, I did not do any change. Everything was in one file. It was transmitted, received, re-transmitted and got accepted. So that is one beautiful way of doing things. And you know, hopefully, that we have to do these things in the future. But still, right now it is mostly by surface mail. Faxes are used only to incorporate changes in manuscripts."

It appears that e-mail is emerging as the primary mode of communication in the first stage of publication, including some cases of collaborative projects. This usage of e-mail seems to vary among different organizations as the state universities in the study which do not have access to e-mail facility seem to lag behind either in terms of quality, or time and of production of knowledge compared to those scientists who use electronic media. This situation creates inequality in science, a digital divide or 'haves' and have- nots may be seen. One of the professors at the IITs had this to say:

"That's the way things are going to be from now onwards. Because a scientists in USA, UK or any developed nation, since they can check e-mail at home also, they can check periodically, and communicate, and they even submit proposals online. There is something called a Fast track in NSF (National Science Foundation) where you submit proposals online, everything in papers you submit online and everything is going to be via computer. It is definitely necessary that, people communicate and modernize the mode of communication.

We have various media of communication. Surface mail, fax and e-mail. Which one do you think is mostly used these days? What do you use most in communicating results in your research?

Yeah! With collaborators, as I wrote in that thing, I use mostly e-mail. That is the fastest and easiest way to communicate with collaborators. Communication in the sense of publishing papers I think. still goes by fax or surface mail. Only recently we started sending by e-mail. They will

> accept but finally even royal society needs a document printed in hard copy. I think days will change I think with the advent of digital signatures and all."

As mentioned earlier, CMC also helps in claiming priority to one's work. A synthetic organic chemist from a national laboratory, remarked that India is technologically backward, India's communication, especially, filing of patents in science compared to the United States. He observes that computer mediated communication provides opportunity to file patents online, helping scientists racing for priority. He said:

> "Faster rate of communication. If you compare Indians with Americans, they file their patents faster because they file their patents online. But in India it is different. By the time I send you the patent which is to be accepted, I lost priority because it's something like an advertisement."

Some of the respondents discussed the role of computer mediated communication technology in triggering a paradigm shift in knowledge acquisition from print journal to Internet related sources.

> "Definitely, it leads to quick publication. You can correspond with some foreign authors. Plus you can have up to date knowledge via Internet. Suppose, previously we have to go and search all the chemical journals. Now with Internet we can get it within a day on a particular topic"

Having seen the role of CMC, scientific communication from the accounts provided by scientists, we shall now see the distribution of e-mail facility, and the frequency of use of and its influence on productivity among the organic chemists. Even though the organic chemists or the state universities recognize the importance of CMC, they did not possess the same.

To quote a young scientist who graduated from a central university and employed in a state university,

> "Here computer is not there. I cant check mail, What to do? I can't help it. I am going to one Dishnet near by. Now-a-days, on internet you get everything. The problem is you get everything on Internet .The main problem here in India is you can't subscribe for all journals. But some of them are on the net. You can print. Bhaskaran in IIT is doing the same thing. If you can e-mail to the author you can ask for reprints and they send also. You can save a lot of time."

6.4 E-Mail and Scientific Communication

In our study, we asked the respondents, if e-mail improved their research productivity. For this subjective question, 83.17% of the respondents gave a positive response. We examined as to what extent e-mail facility contributes to productivity of the organic chemists in the study. We asked the respondents whether e-mail made any critical difference to their research, what purposes does e-mail serve etc., we categorized the responses, into the following categories:

- Economical alternative
- Time
- Distance
- Knowledge
- Collaborators
- Publication process

6.4.1. E-Mail: An Economical Alternative

In the earlier chapter, we gave an account of how e-mail facility is procured and maintained. Except state universities, where scientists access e-mail at non-academic places, at all other organizations in the study, the e-mail facility is provided by the organization to the scientists and the latter do not incur any expenditure on the same.

Respondents from these organizations mentioned that e-mail facility is inexpensive and helps 'free' exchange of information. Thus computer-mediated communication seems to emerge as an economic alternative to traditional modes of communication in science, like telephone, fax and surface mail, which are more expensive and more time consuming. A scientist from the Central University said:

> "e-mail has really transformed the way we communicate. For ex: in a single day, I had exchanged about 5 messages with one person (student/collaborator). Such exchanges of e-mail messages are very useful in communicating with others who are at a distant place. This inexpensive mode of communication thus has helped researchers very much. I can not think of making 5 phone calls in a day to someone in Germany."

6.4.2 E-Mail and Distance

E-mail resulted in the death of distance. This instant communication mode facilitates instant interaction with anyone anywhere in the world… to quote a solid state structural chemist from central university said:

> "I can sit anywhere in the world and do my research"

6.4.3 E-Mail and Time

Traditionally, correspondence with peers in the scientific community used to conserve more time. The e-mail communication facility has almost zeroed, the time of correspondence, thereby providing faster access to knowledge, exchange of information, related to research, funding and collaboration. Etc., One of the scientists mentioned:

> "It has been easy to contact researchers in the same area. In addition, progress of the refereeing process of submitted papers is intimated (to me) fast.
> Easy to search literature
> Takes less time to obtain a reprint of an article
> Takes much less time to contact a person
> Can't imagine without having the e-mail facility could prepare a joint research proposal (collaborative in just one day with the collaborator stationed in Germany. This is made possible because we could transmit several drafts of the proposals in few hours".

6.4.4 E-Mail and Knowledge Needs

The advent of electronic mail communication brought about a new dimension to acquisition of information/knowledge in science. The Internet, World Wide Web and electronic journals facilitate the availability of abstract of a research publication along with the e-mail address of the authors. Those interested in the same in the first place, access the abstract online and also correspond with the author immediately. It seems to be extremely useful accessing journals and scientific works which are not available in the print medium in the academic libraries.

6.4.5 E-Mail and Collaboration

Scientists involved in the research enterprise, as a norm are in cooperative relation to one another in terms of exchange of information, sharing of ideas, and technological and other modes of interdependence. They, thus communicate with each other on regular basis, in this context, communication emerges as an essential process within a collaborative front.

Collaborative projects undertaken by scientists from different departments, organizations and countries, require constant interaction among themselves. Traditional modes of communication used to consume lot time and money. E-mail seems to have provided an alternative by being instant and inexpensive. We observe that the majority of the scientists except those from state universities and those who are involved in the intra-institutional collaboration shifted to e-mail as the most frequently used mode of communication with their collaborators. We examined if there is any relationship between volume of communication and number of collaborations. (See Fig. 6.4.5)

As mentioned above, volume of communication is measured in terms of total number of incoming and out going messages in a week.

In our study we obtained data about the number of collaborations each scientific group is involved in. We in turn have inquired into with what frequency do they communicate through e-mail with collaborators.

We hypothesize that, both communication and collaboration influence the productivity of organic chemists. We wished to examine as to what extent communication aids in collaborative research. (See Table.6.4.5)

We observed that e-mail seems to emerge as a significant mode of communication in science. Hence we considered the frequency of the use of e-mail with respect to the research as one of the important independent variables to explain the variability of productivity among organic chemists in India along with other variables.

Table 6.4.5: Volume of communication and number of collaborations

VOLUME OF ELECTRONIC COMMUNICATION	NUMBER OF COLLABORATIONS					
	0	1	2	3	4 above	TOTAL
0	2	1	-	-	-	3
1-50	8	5	5	5	4	27
51-100	-	4	2	1	1	8
100-150	2	1	-	3	7	13
150 AND >	-	-	-	1	1	2
TOTAL	12	11	7	10	13	53

Our study revealed that 22.64% of the respondents were not involved in any collaborative research at the time of the study. Three of the respondents (5.66%) were found to have not been involved in collaborative research and rated 0 in terms of volume of electronic mail communication. Twenty-eight respondents (52.83%) have replied for being involved in electronic communication and up to 4 collaborations. Thirteen respondents (24.52%) communicated extensively in electronic mode, as well were involved in more than 4 collaborations.

Ten respondents were found involved in no collaboration despite having rated high on volume of electronic communication. The graphical representation (Fig.6.5.5) shows the trend. It shows that greater the number of collaborations, the greater is the volume of communication.

Fig.6.4.5 Volume of Communication and Number of Collaborations

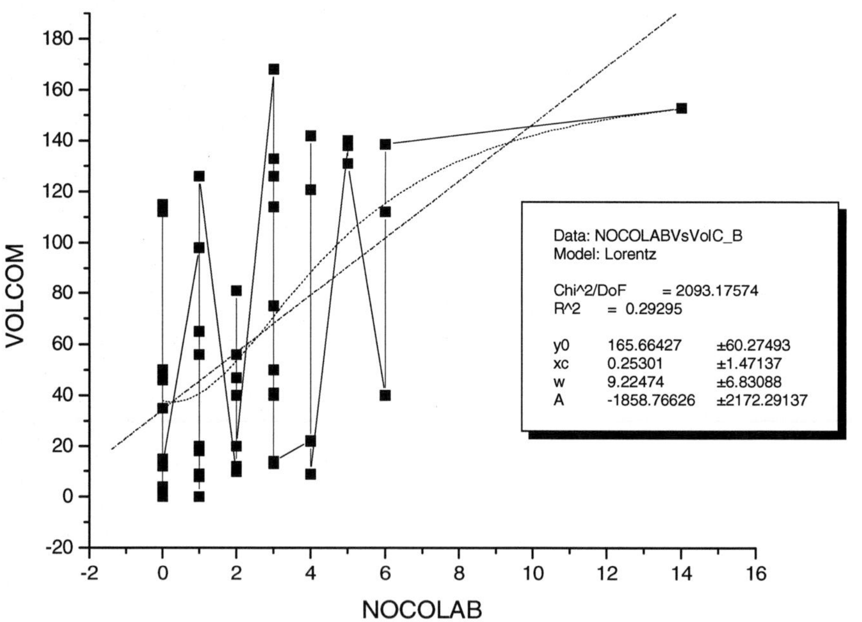

6.4.6 Publication Process

Before the advent of e-mail, the mode of communication itself used to be a determinant of hierarchy among scientists. Experts in a discipline were not easily accessible. The advent of e-mail facility resulted in easy and instant access to these knowledge gurus wherever they are.

E-mail, though a recent innovation, seems to be emerging as the most predominant mode of communication among collaborators. Communication in some cases seems to be through surface mail or fax. With respect to sending the manuscripts, e-mail has speeded up the process. One of the scientists said:

> "Absolutely it has, because it is not only from my side and editor's side because publishing also has editors and referees. It has to go through many cycles and circles that makes things faster."

Another scientist said that e-mail helps in getting publications faster. Some scientists feel that e-mail does help extending collaboration. Some of them said:

> "Access to e-mail and Internet; does it have scope to extend the collaboration? For instance number of collaborators.
>
> In certain subjects, yes. May not be in experimental science. Theory and other things."

Knowledge claims in science are made through an organized system of publication. The publication process involved human beings as actors and technology as mode of communication. Experts in a discipline act as referees, i.e., status judges to evaluate the work of their peers through an institutionalized mechanism called peer-reviewing process. The entire cycle of communication involved different stages such as sending the manuscripts, forwarding the same to referees etc.

At the stages of communication of one's research, a scientist relies on media of communication like surface-mail and fax. These modes involved expending lot of time. E-mail facility seems to reduce the time lag and accelerate the publication process. Says a scientist,

> "My last publication was sent through e-mail and got peer-reviewed by e-mail and Galley proof was corrected through e-mail. The whole process took only 1.5 months, which would have taken about 6 months by regular post".

> "I could publish my work faster when I travel abroad, my students can communicate with me very well. Infact, if there is any doubt, on spectral data, they can scan and send via e-mail and I can opine on that".

E-mail thus, according to some respondents quickened the publication process. We feel that, scientists who don't have access to this CMC technology and among those who make less frequent use of e-mail seem to be fall behind in terms of number publications or lead to productivity differentials.

The entire process of communication related to publication takes certain amount of time. Some journals do not accept some publications. In such a case, says a scientist, e-mail provides an instant opportunity to send the same to another journal for publication. Thus e-mail seems to have consequences for rejection rates in science. To quote one of the scientists:

> "It is fast, the decisions regarding a paper's acceptability/rejection, fast so one can send to another journal without delay".

6.5 Age and Volume of Communication

The study aimed at identifying the adaptability of scientists to electronic mail according to age. From our study we observed that scientists' adaptability to e-mail mode of communication varies according to the age group. (See Table. 6.5) We found that scientists in the age group 40-49 (64.15%) are highly electronically connected. Those below 40 years of age have registered in the category of 1-50 messages a week on an average.

Table 6.5: Age and Volume of Communication

VOLUME OF COMMUICATION	AGE GROUPS OF RESPONDENTS						
	25-29	34-39	40-44	45-49	50-55	55-59	TOTAL
0	-	-	-	-	1	2	3
1-50	1	5	7	9	5	-	27
51-100	-	1	4	2	1	-	8
101-150	-	-	9	2	1	1	13
151-200	-	-	-	1	1	-	2
TOTAL	**1**	**6**	**20**	**14**	**9**	**3**	**53**

The majority of 20 scientists in the age group 40-44 (37.73%) were identified, who had high volume of communication, followed by those in the age group 45-49 (26.41%). In the category of above 50 years of age, 9 scientists were found and 3 scientists in the age group of 50 years, one from CSIR laboratory and 2 from State University never communicated at all. (See table 6.5). It appears that, the communicative needs organic chemists in the study start at an age of 34 and reach the peak among scientists between 40-49 and then slowly declines after 55 years. (See Fig.6.5)

Young scientists seem to, take time to socialize into organizational and human environment in which they are located. Afterwards they start establishing contacts to exchange information on topics related to research, conferences, funding, project proposals, manuscripts to editors. This trend

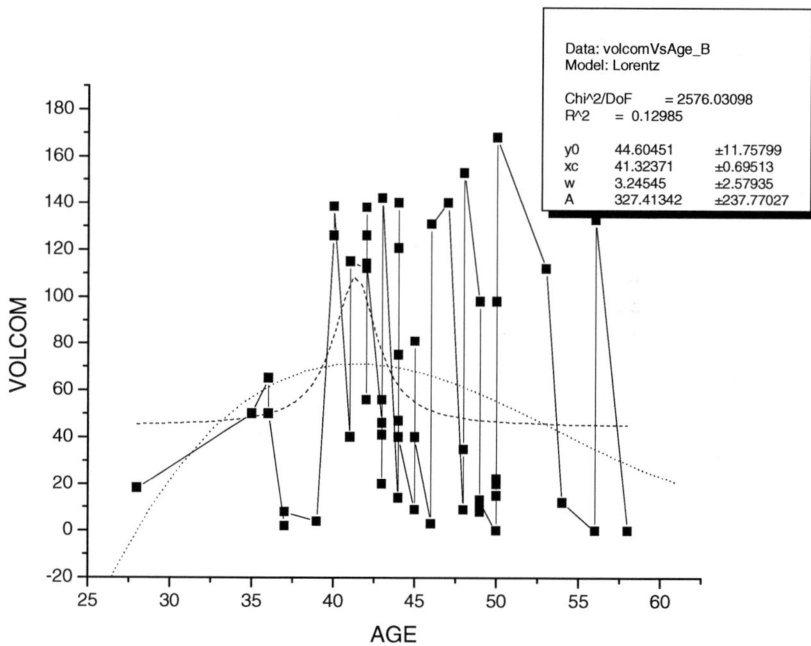

Fig.6.5 Age and Volume of Communication

seem to reach between 40-49 years and it appears that scientists slowly take upon themselves administrative role after the 50th year. (See Fig.6.5)

6.6 Volume of Communication and Productivity

The study considered the total number of e-mail messages sent and received in a week with respect to research as volume of communication. We intended to find out the impact of this volume of electronic mail communication on the productivity level of organic chemists. I considered the differential frequency of use of e-mail. I found that majority of the scientists 50 out of 53 use e-mail for research related communication and are productive, while those who do not use the same seem to appear low in productivity. (See Table 6.6)

Table.6.6: Volume of communication and productivity

VOLUME OF COMMUNICATION	PRODUCTIVITY							
	0	20-29	10-19	20-29	30-39	40-49	50 and above	TOTAL
0	-	2	-	-	-	1	-	3
1-49	3	5	4	9	2	2	2	27
50-100	-	2	2	2	1	1	-	8
101-150	-	-	-	4	4	3	2	13
150>	-	-	-	-	1	-	1	2
TOTAL	3	9	6	15	8	7	5	53

Twenty seven of the respondents fall in the category of 1-50 messages per week of and have been highly productive. Eight of the respondents fall in the category of those whose volume of communication is 50-100 messages having published in all levels of productivity from 1-49 publications. In the case of thirteen respondents, the volume is 100-150 messages per week. Their publications range from a minimum of 20 to above 50.

Two Professors, one from solid state structural chemistry from a central university and another from an IIT, rated high on the frequency of volume of communication, that is, 150 and above published above 50 papers and around 40 papers respectively. The professor when asked about the role of e-mail in research, remarked:

> "Yeah, you can sit anywhere in the world and do your research, **e-mail shrinks distances, across geographical boundaries**, you can sit any where in the world and do your research, just as you are sitting in front of me and discussing this. **I can be in touch with anyone anytime from any where and do my research"**

It helps discuss ideas with any one, anywhere in the world, sending publications, it **opens up your mind**. E-mail **removes psychological barriers, it removes the element of secrecy that plagues Indian science,** and in fact it has changed our life."

Fig 6.6 Volume of Communication and Productivity

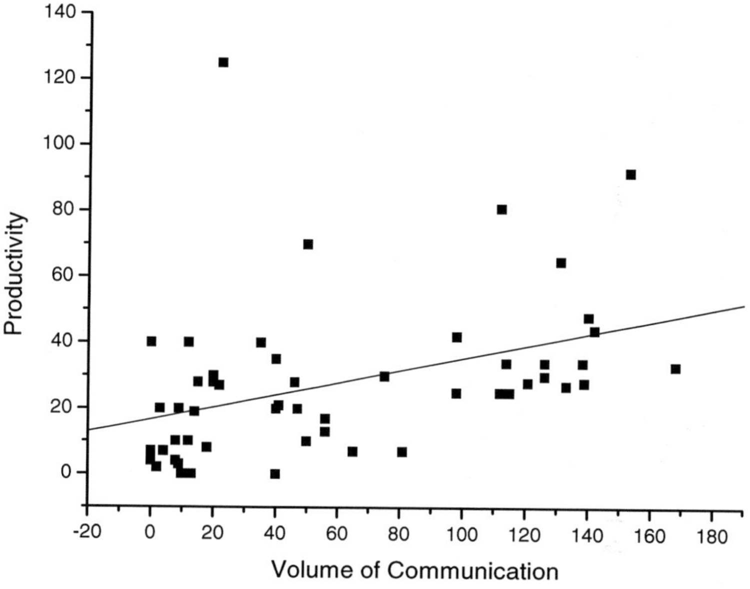

From the figure it appears that the productivity levels increase as the volume of communication increases. (See Fig. 6.6) We may conclude that there is a positive association between volume of electronic communication and productivity.

SECTION-II

6.7 Collaboration in Science

In our study we considered collaboration as a social process in research in which two or more individual scientists from the same organization, different organizations, and different countries are involved.

We asked the scientists what according to them is collaboration, why do Indian scientists collaborate, what is the basis of collaboration in India, why is it that Indian scientists collaborate with scientists abroad? What are the advantages of being involved in collaborative research and disadvantages as well?

On the basis of the responses of the scientists during interviews, it emerged that the following needs motivate scientists to collaborate.

1. Knowledge needs
2. Needs for coping with inter-disciplinarity
3. Needs for expertise
4. Technological needs
5. Need for resources

We also observed that the need structure is not a watertight compartmentalization, but a sphere of overlapping circles

6.7.1 Knowledge Needs

Modern science is an ever-expanding enterprise. The processes of industrialization and globalization and increasing questions on life and causes for existence and survival demanded and as well as resulted in expansion of knowledge, we presume results in information overload, and diversity. Time being as essential commodity, scientists seem to cooperate with each other for want of information to be exchanged.

There is another group of scientists, the members of which are oriented toward the growth of the discipline as advancement of science. They perform research for the sake of disciplinary growth and in the process collaborate with other scientists. To quote a scientist from the central university,

> "In this modern era collaboration is a must. Of course you know so much is there in my own field, I cannot cope up with myself okay. It is better to collaborate. I collaborate very extensively with people. That is the only way, because I for sure know what I know. It is better if you want to be productive in the sense that I don't mean number of publications, but for advancement of science, it is better in modern days to collaborate. Especially when we work in very inter-disciplinary fields"

6.7.2 Needs for Inter-Disciplinarity

Emergence of new branches of knowledge draw inputs from different disciplines blurring individual boundaries. The inter-disciplinary nature of the emerging research areas requires the integration of theories and methods of different disciplines. The aspects of knowledge and technology are available in a relative degree among individuals and institutions respectively. Thus access to information and technology cannot be concentrated for want of economic reasons and scientists tend to enter collaborative research. To quote an organic chemist from a national laboratory,

> "I develop an expertise. I require other expert to be able to help for meaningful project completion. I collaborate with chemical engineering department. Science is science, and it is not compartmentalization like chemistry, physics, and biology."

6.7.3 Needs for Expertise

In research, in terms of the specialization, and expertise, scientists varying in different disciplines tend to possess different types and degrees of expertise depending on the proven use of duration particular technologies. As it is impossible for all scientists to master all theories, methods and experimental skills, expertise in general gets localized in some cases individual specific. The expertise could also be of information in both the cases, scientists require- expertise from others and such a situation necessitates them to enter into a collaborative research. To quote a scientist from a national laboratory,

> "I develop an expertise. I require other expert to be able to help for meaningful project completion. I collaborate with chemical engineering department. Science is science, and it is not compartmentalization like chemistry, physics, and biology."

Also says another synthetic organic chemist from the deemed university,

> "To add up to expertise in different fields, like for example, Crystallography and organic chemistry."

6.7.4 Technological Needs

Different organizations have different kinds of infrastructure related to research through different modes of acquisition. They also vary in quantitative and qualitative terms, in terms of the possession of technology. Those who lack seek from those who have. Thus need for technology both for short and long terms needs, results in scientist enter collaborative front. To quote a scientist from a state university,

> "Just now we had to fax a recommendation letter to our student. We don't have it in the university. I am waiting for a fellow. We have to send it from a Industry. We don't have money to do this.
>
> Collaboration is a must now a days. We being a poor country, we have to collaborate with INDUSTRY. We can not do research by Governement Funding. If any body collaborates with industry he gets money.... Financial assistance. At present there are 2 or 3 parties in MOU stage.
>
> That is known.. we hardly have any instruments....
>
> Then how do you manage your research?
> We go to IICT, we use their NMR-specturm.....
> Magentic Resonace.. and Other excellent facilities..."

6.7.5 Need for Resources

To produce knowledge, scientists need, inputs in terms of journals, print and electronic media and a variety of infra-structure and support for acquiring technology related to communication like, computers, internet, etc. Different organizations have different levels of financial endowments. This is reflected in the infrastructure support they provide to the scientists employed by them. Some scientists acquire some input mobilized through the funding provided by government funding agencies, industry. We begin to see a change in the Indian science from a stage of total dependence on the state agencies for research funds, scientists are exploring the possibilities of funding from industry. One of the scientists remarked:

> *"Collaboration is a must now a days. We being a poor country, we have to collaborate with INDUSTRY. We can not do research by Government Funding. If any body collaborates with industry he gets money.... Financial assistance. At present there are 2 or 3 parties in MOU stage.*
>
> We have a project given by Andhra Pradesh Government and Netherlands Governement. The budget given is spent in 2 fractions. As the project involves study on insects 75% of the money goes on the field trips. Only 25% is left for expenditure in the laboratory."

6.8 Age and Collaboration

We examined whether scientists' propensity to collaborate varies with the age of the scientists in the study. From the data collected we observed that scientists in the age between 40 and 49 years seem to collaborate more. (See table.6.8)

Table 6.8: Age groups and number of collaborations

AGE	NUMBER OF COLLABORATIONS					
	0	1	2	3	4 and above	TOTAL
25-29	-	1	-	-	-	1
34-39	3	2	-	1	-	6
49-44	3	3	2	6	6	20
45-49	2	3	4	1	4	14
50-54	3	1	1	1	3	9
55-59	1	1	-	1	-	3
TOTAL	**12**	**11**	**7**	**10**	**13**	**53**

The involvement in collaboration is at its highest from the 40th year being more productive, and from the 50th year on wards the tendency to collaborate seems to decline. A relatively small number of respondents, 9 i.e, (13.20%), in the age group 50-54 years, category were involved in more than one collaborative project with 1/3 rd of them having no collaboration at all. From the graph (See Fig 6.9) none among the scientists in the age group 55-59, a meager 5.66% in our study has more than 2 collaborations. Thirty four scientists in the age group 40-49 have active collaboration with 17 of them being involved in more than 3 collaborative projects, and 6 of them being involved in one and 2 collaborative research ventures.

Young Scientists from recruitment take time to establish contacts, naturalize themselves to the organizational context, then initiate and enter into collaboration. Hence relatively less number of young scientists are involved in research collaboration. The data shows that collaboration slowly begins and reaches a peak and then declines.

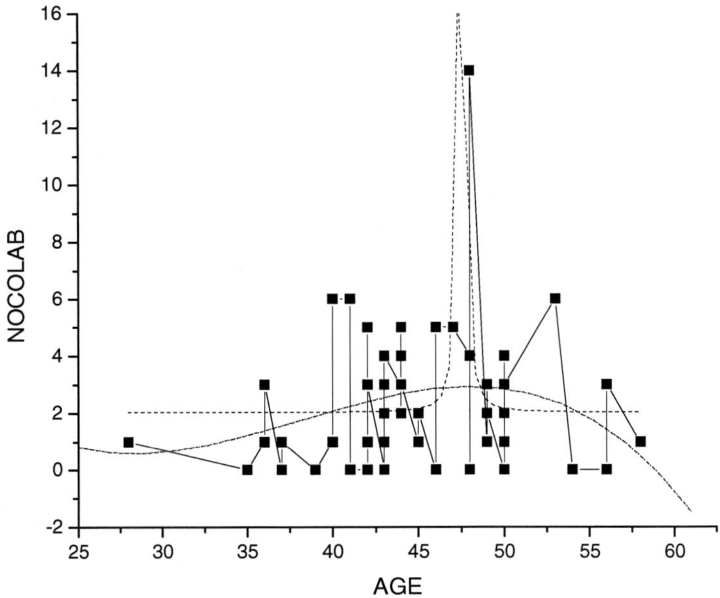

Figure 6.8: Age and collaborations

6.9 Similarity of Research Area and Collaboration

In our study, we observed that who continue in the same are of research after the doctoral program, are involved more in collaborative research and more number of collaborations. 23 of the respondents continued to be in the same area of research of which 13 have more than 3 collaborations and 7 of them have been involved in nearly 2 collaborations. While, 3 are not involved in any collaborative research. Of which 2 are from a state university and one from a national laboratory.

Of the 8 scientists who have almost completely changed the area of research, 4 respondents are involved in more than 3 collaborations and other 4 in 2 collaborations. 5 have no collaboration. Of the 5, two are from state university with 54 and 56 years of age, one from central university aged50, and two from IITs aged 35 and 41.(See Table.6.9)

Table 6.9: Similarity of research area and number of collaborations

SIMILARITY OF RESEARCH AREA	NUMBER OF COLLABORATIONS					
	0	1	2	3	4 and above	TOTAL
TO A VERY SMALL EXTENT	5	3	1	1	3	13
TO A SMALL EXTENT	1	2	2	-	1	6
TO A LARGE EXTENT	3	1	2	3	2	11
TO A VERY LARGE EXTENT	3	5	2	6	7	23
TOTAL	12	11	7	10	13	53

Thirteen scientists are in the fringe wherein the degree of similarity is medium. Of the thirty-four scientists fall under category-I, similarity to 18 respondents had more than 3 collaborations and 10 scientists up to 2 collaborations, while 6 did not have collaboration. Nineteen respondents, fall in category II is the similarity to a small extent. The majority in the group reported that they had three collaborations, 8 respondents are involved in up to 2 collaborations. Six scientists had no collaborations at the time of the study.

It appears that, the larger the similarity of area of research with that of the doctoral thesis and the area of interest at the time of the study, the more is the opportunity to collaborate. Researchers working in a similar area over a long period would have opportunities to establish communication and network relations. Similarly, the smaller the degree of similarity of the area of research, the less is the scope to collaborate. Perhaps, the scientists who shifted to a new area were in the process of establishing network relations in the new area.

6.10 Basis of Collaboration

In the study, we attempted to examine on what basis collaboration among scientists occurs in the Indian context. It is observed that in India, collaborative research is initiated based on personal contacts and personalities rather than on the basis of complementary expertise.

To quote a professor from a state university,

At *personal* level, two people from different fields, work. There may be positive or negative results. But a systematic survey happens. For example, In a collaborative effort that involves human physiology and medicinal properties, you may never know, by taking a lean compound, studying the structure, if you synthesize further, it could become effective research.

Another scientist from the deemed university remarked:

"Whereas abroad, collaborations interactions are limited to work, than personalities, whereas in India, it depends on personalities".

Asked to explain why collaboration takes place on the basis of personality in the Indian context, scientist from the central university said:

It is very primitive. In India it is all personality based. I don't know. May be it is due to 'training'. In commas 'training'. I don't say it's personal egos but we have been built-up like that. 'That don't tell this thing to somebody else unless you really have it in your hand'. That kind of attitude is a sheer attitude problem than anything else.

We encountered scientists who also mentioned that collaboration is very primitive and not yet a mature process. They contend that, in the Indian context, neither institutional structures, nor consortia exist as a prescriptive-normative apparatus underlying collaborative research. Says a scientist from IIT:

"I don't think we are mature enough, because, we always talk about collaboration in terms of individuals saying so and so is doing an excellent job in promoting, international collaboration.

If I have collaboration with someone in India and publish papers together, it doesn't matter, how good the paper is or what good for the sake of number crunching. It becomes only half. Ok. So we haven't become matured enough for collaboration. The collaboration more or less it is looked at is still like the number crunching dinosaur sitting at the administration."

We inquired whether collaboration with foreign scientists is emerging as a trend, if yes what could be the factors? Scientists emphasized upon the fact that since India is a developing country, with characteristic constraints, lags behind in terms of many resources from that of developing countries, and hence, Indian scientists collaborate with foreign scientists, among other factors for superior- infrastructure. To quote another scientist from an IIT:

> "You see I can not deny that they have superior infrastructure. We are still a developing country, we have not become a developed nation. So by shutting yourself off from the outside world, you are going to be I mean you know some country like Iran or Iraq or some thing. But then I don't think it's a good thing for India. We have to coalesce."

Another scientist from deemed university, says collaborative research with foreign scientists, institutions brings out monetary gains in terms of funding and relations, meaning networking with people from different countries, networking is an essential aspect in collaborative research forming a skeletal structure facilitating flow of information among different actors. Same scientists also emphasize on collaboration leading to good science, good science in the sense of quality of research. To quote another scientist from the deemed university:

> "Why does one collaborate with foreigners?
> That is for
> 1. Money 2. Relations and 3. To do good science".

Lack of or absence of significant Industrial R&D thrust in India, has also been quoted by the scientist for the lack of adequate support for research among scientific organizations. They claim this is also as one of the factors for Indians to collaborate with foreign scientists. A scientist from a national laboratory said:

> "In the absence of Indian industry not coming forward to spend on R&D at research institutions there is no other way for scientists at Indian institutions to look at the foreign collaboration and take job works."

6.11 Collaboration and Productivity

I classified scientists into groups basing on the number of collaborations they were involved in at the time of the study, which ranged from 0 collaborations to 4 collaborations and above. We categorized productivity levels in terms of the number of publications ranging from '0' to '50' and above divided. Thus, 7 classes with 10 publications in each of the class interval was evolved (See Table 6.11).

Table.6.11: Number of collaborations and productivity

NO COLLAB	PRODUCTIVITY							
	0	0-9	11-19	20-29	30-39	40-49	50 and above	TOTAL
0	-	4	2	5	-	1	-	12
1	-	4	2	1	2	2	-	11
2	2	1	1	2	-	1	-	7
3	1	-	1	2	5	-	1	10
4>	-	-	-	5	1	3	4	13
TOTAL	3	9	6	15	8	7	5	53

NOCOLLAB: Number of Collaborations

In the study, 12 out of 53 respondents have not been involved in any kind of collaboration, of which 11 have had published up to a maximum of 30 and only one synthetic organic chemist from central university has published over 40 articles.

The majority of 41 scientists have been involved in collaborative research. The number of collaborative projects ranged from 1 to '4 and above'. There seems to be a positive association between number of collaborations involved and the productivity level. (See Fig. 6.11). Those involved in a maximum of 2 collaborations (13.20 %) have published up to a maximum number of 49 publications. Two scientists in this category, in the age group 45-49 from applied division of national laboratories have rated at 0 productivity measured in terms of research papers. The category of scientists involved in 3 collaborations (18.87 %) seems to be more productive, with one of them having more than 50 publications to his credit. In this category we found one scientist, aged 49, from the applied division of a national laboratory, despite having been involved in 3 collaborations, to have no publications to his credit. In the group of scientists who have 4 and above number of collaborations (24.53 %) the productivity level starts from a minimum of 20 publications to a maximum of 4 scientists having published more than 50.

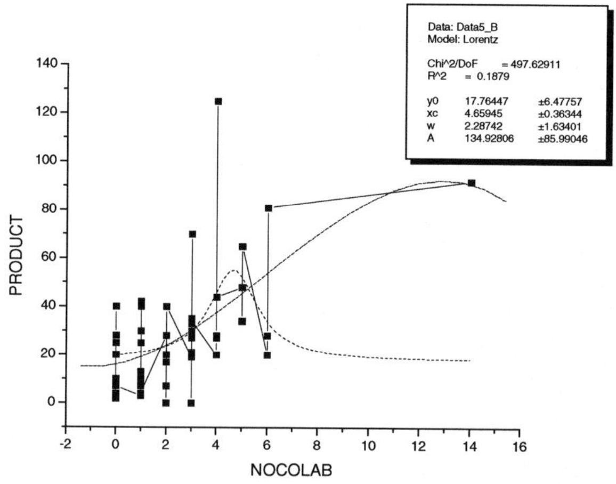

Figure 6.11 Number of collaborations and productivity

Thus it seems that, increase in the number of collaborations is in positive association with increase on productivity levels.

6.12 Constraints in Collaboration

We questioned the scientists on the constraints in collaborative research. We categorized the responses into four classes,

- Credit Sharing
- Economic- international exchange programs, financial constraints
- National Relevance
- Personal collaboration and credit sharing.

In general, it is assumed that a collaborative research effort manifests in terms of publications or in some cases, patents. The publications allocate credit to scientists by identifying them in terms of first and co-authorship. Sharing of credit seems to be a problem from the point of view of evaluation of the scientists for various purposes like promotion and reward, by the non-scientific administrators. Scientists question the ignorance of the administrators about the quantity and quality aspects of collaborative research. A scientist from IIT said,

> "I write a publication that's another evaluation. If I publish 5 other names in a reputed journal like Angkorvodt, or something, it really did not matter because what these administration guys do is ask me to do is to underline these students' names. Okay, then lets assume that these students are not principal administrators and I have 4 names. Like I had one publication in physics department on C60 with my name and another professor from physics department and 2 other scientists from kalpakam who have made the C60 and we have the done the purification of C60 and third name being optics measured by somebody in physics department. He published this paper. It is a wonderful paper and it has been quoted by many people but when I write this publication for any kind of evaluation here, it's the students name which is underlined. There, there are 2 students names, are from my group, and other from physics department and totally there are six authors names there. So there are 6 names one from physics, one from here and 2 from kalpakam. So my contribution in this paper is ¼. Even though I feel it is a very good paper, which is standard and is supposed to be cited by others, in that field, the administration thinks there are 2 students, and there are 4 different guides, who are not students so, lets assume that their contribution is same from each one of them. So my contribution is only $1/4^{th}$. Rationalized. So this happens in terms if I have a collaboration with someone in India and publish papers together I t does not matter, how good the paper is or what good for the sake of number crunching. It becomes only half. Ok. So we haven't become matured enough for collaboration. The collaboration more or less it looked like it is looked at is still like the number crunching dinosaur sitting at the administration. So we have to do something about it".

Credit shared out of collaborative research if based on personalities, is supposed to lead to inequality and the scientists reinforce hierarchy in terms of authorship. Says a Professor from deemed university,

> "Collaboration leads to problems- like conflict of interests, hierarchy in publications"

In some instances, some scientists complained of being forced to give equal credit irrespective of their extent of individual's contribution to collaborative venture. In the second place, we examine the financial constraints. Scientists complain of lower amount of financial assistance to those who visit foreign countries under international exchange programs. They feel that the Indian funding agencies sanction the financial assistance according to norms institutionalized at least 10 years ago irrespective of the present cost of living abroad, thereby putting scientists to a great deal of inconvenience. A scientist from IIT remarked,

> " Constraints.... 1. Of course monetary, because all said and done, many of these collaboration programs in Japan are contracts written between two agencies like Indian national science academy, royal society of chemistry and so, lets say something that is done may be once in twenty years or so. So, they fix some monetary scale on that and it is never revised. Recently, I went to London. Or may be 4 or 5 years back on one of these exchange programs. And you know what I was given? I was given 10 pounds a day. Which is very meager you know. May be you can live on it. But not as a scientist or somebody like a visiting professor. But this, they write to the secretary of Royal society or this I write to the secretary of INSA. They said, there is little that they can do as it was already fixed in 1980. You see, this is what it was fixed in 1980. 1980 To 2000, 20 years of difference. I mean, may be they have revised it now. But this was what they sent me on, as a fact. Some of these things we have to look into. You cannot still hang on to the old system because it is typical. I am not saying that you spend so much money but something on par with others. I see scientists visiting from abroad to here. And they are very good you know. They go around, spend on this and they can reimburse it later and so on. Ours is not flexible at all in our system. We should allow for that change of time. We should say that things should be more flexible and then incorporate international expenses."

Finally, one scientist questioned the relevance of collaborating with foreigners. He observes that Indian scientists by obtaining from foreign bodies have a normative compulsion to solve research problems relevant to foreign context than those related to Indian context. This scientist from a national laboratory said:

> "Collaborative research has no problems excepting that.... Indian scientists working in national laboratories, are solving problems unrelated to Indian demands and aspirations."

6.13 Summary of Findings

The respondents identified that journals are still sources of ideas and there is differential access to print the emerging electronic journals. We observed that in science conferences are communication events. From the data we observed that conferences facilitate exchange of information, exposure to new fields and frontier areas of research, initiate collaborative research. Conferences create opportunity for establishing priority claims in science. Conferences propel balanced reciprocal economic relations among scientists in terms of exchange of funding and financial sources.

Conferences facilitate social interaction among scientists, initiate, invisible and visible networks in the scientific community. Conferences also enroll new actors into the scientific networks. Conferences are occasions, wherein young doctoral students receive professional socialization, enroll themselves into scientific networks for economic, social and career opportunities in addition to meeting their knowledge needs.

All over the world, E-mail seems to play an important role in communication at different stages of production of knowledge in science.

CMC facilitates quick publication and symbolizes a paradigm shift in accessing, diffusion, production and certification of knowledge. CMC aids in priority claims among scientists by proving a faster means of communication in science. In our study, scientists claimed that those who don't have access to computer mediated communication technology (CMC) tend to lag behind in terms of productivity.

CMC technology is absent in the state universities as an organizational facility. Scientists from the state universities are excluded from the mainstream scientific community in India, comprising of scientists from the central universities, IITs, National Laboratories and deemed universities, because they are not part of the 'digital communication' world. In collaborative research e-mail seems to be a predominant mode of communication.

Among the respondents, 83.17% perceive that, email increases productivity. E-mail seems to be an inexpensive and alternative medium of communication among Indian organic chemists. Except scientists from state universities, majority of the respondents seems to rely upon the email facility to interact with collaborators. Thus, email seems to emerge as an essential medium of communication that supports an interactive network of scientists in a collaborative front.

In the sample under study, 52.83% of the respondents belong to high frequency users of electronic communication in maximum number of collaborative projects. Scientists in the age group 40-49 (64.15%), rank high in terms of usage of e-mail communication. It appears that the communicative needs related to research and productivity are high during this period. Within this category of scientists, those in the age group of 40-44 are high in terms of their volume of communication (37.73%) followed by those in the age group 45-49 (26.41%).

I found that almost 98% of the respondents use email for research communication. We observed that volume of communication may be divided into four categories: very low: 1-50 messages per week, low 51-100 messages; medium 101-150 messages and high 151- and above. It appears that volume of electronic communication plays an important role in contributing to productivity of the scientists.

It is observed that in the Indian context, collaboration is based on personality rather than being limited to work. Collaborative research is still primitive in the Indian context, is not yet a mature process due to the absence of institutional-structures consortia as prescriptive normative apparatus. The data reveal that scientists collaborate among themselves to fulfill, technological, economic, knowledge needs and needs of coping with growing inter-disciplinarity. Scientists in the 40-49 years age group tend to collaborate more.

The degree of similarity of research area seems to be positively associated with number of collaborative projects. The greater the degree of similarity of the research area, the greater is the scope for collaborative research. Technological needs seem to be factors responsible for entering into collaborative relation with scientists from abroad. Economic considerations also seem to be factors responsible for entering into collaborative relation with scientists from abroad.

I find that collaborative research results in; 1.Increase in the number of publications; 2.Speed of publication process; and 3. New areas of research/discipline on prolonged pursuit.

It is observed that there is a positive association between number of collaborative projects and the productivity levels. We observed that those who are not involved in collaborative research fall in the category of low productivity.

In the Indian context, collaborative publications under the purview of bureaucratic evaluation for career advancement leads to the treatment of all the collaborators on an equal basis neglecting the major contributor with higher recognition for the work. Personality based collaboration leads to hierarchy in publications. Collaborative ventures like exchange programs, being insufficiently funded, act as demotivators in Indian Science.

CHAPTER SEVEN

VARABILITY IN PRODUCTIVITY: A STATISTICAL EXPLANATION

Introduction

In chapter 4, I presented the profile of organic chemists in our study in the form of frequency tables. In chapters 5 and 6, I attempted to examine the association between productivity on the one hand and a set of independent variables like age, duration of employment, duration of postdoctoral training in India and abroad, Doctoral affiliation, Volume of communication, average size of the research group, similarity of research area, and number of collaborations through contingency analysis. The contingency analysis showed a positive association between productivity and the above mentioned independent variables.

7.1 Empirical Analysis of Factors influencing Productivity

In this section I shall analyze the relative contribution of the factors to the productivity of organic chemists by using multiple regression analysis. The dependent variable used for this analysis is Productivity of the scientists measured in terms of the number of publications they had in refereed journals during the period of five years (1996-2000), The independent variables are: duration of association of the scientists with the present organization, age of the Scientist, average size of the research group in the last five years; number of collaborative projects at the time of the study; duration of the postdoctoral study abroad; duration of postdoctoral study in India and volume of electronic communication; measured in terms of outgoing and incoming message over a one week period.

Linear Regression Analysis is a technique used to estimate the value of one quantitative variable by considering its relationship with one or more other quantitative variables (Pfaffenberger and Patterson, 1987)

The general form of the multiple linear regression model is:

$$\mathbf{Y=\beta+\beta_1X_1+\beta 2X_2+\ldots+\beta n\ Xn+\in}$$

Where,

- **Y-Dependent variable corresponding to X_1, X2,……X_n**
- **β,β_1, β_2……..β_n- parameters in the model**
- **X_1,X_2,……. Xn- exogenous variables and**
- **$\in$-Random error term**

The parameter 'β' determines the level of the line fit i.e., the distance of the line above or below the origin. The parameter 'β1, β2……βn' determines the slope of the line that is the changes in Y per unit change in X.

As mentioned above, in our analysis, we have taken the productivity as a dependent variable and remaining variables as independent variables.

The multiple regression model is as follows:

PROD=
a+**a_1**AGE+**a_2**GPSIZ+**a_3**NOCOLB+**a_4**DURATION+**a_5**PDFADU+**a_6**PDFIDU+**a_7**VOLCOM+E

Where,

a is a constant term
PROD- Productivity of the scientists i.e., the no. Of publications they had in refereed journals
AGE- Age of the Scientist
GPSIZE- Average size of the research group in the last five years
NOCOLAB-Number of Collaborations
DURATION-Duration of association with the present organization
PDFADU-Duration of the postdoctoral study abroad
PDFIDU-Duration of the postdoctoral study in India
VOLCOM- Volume of Communication by scientists
a_1…..a_2…..a_3 …a_6 are the coefficients of independent variables

To analyze the relation among the variables we use correlation matrix. This matrix analysis is used to find out the interrelationship between the explanatory variables on one hand and the dependent variable on the other.

The coefficient of correlation measures the strength (expressed by a number ranging from –1 to +1) and direction (positive or negative of the relation between a set of variables As the reliability estimate depends upon the closeness of the relationship, it is imperative that utmost, care is taken while interpreting the value of coefficient of correlation, otherwise fallacious conclusions may be drawn.

*If r=+1 it means there is a perfect positive relationship between the variables.

*If r=-1 it means there is a perfect negative relation between the variables.

*.If r=o it means there is no relation between the variables, i.e., variables are uncorrelated.

* The closer r is to +1 or –1, the closer the relation between the variables and the closer r is to 0, the less close is the relationship.

Where, 'r' is the correlation coefficient.

7.2 Correlation Analysis

The correlation matrix of the factors determinant of productivity is given in Table 7.2. This matrix analysis is used to find out the interrelationship between the explanatory variables such as Age of the respondents, Average size of the research group in the previous five years total number of collaborative projects the scientific group is involved in, Duration at the organization in which the scientist is employed, the duration of postdoctoral study abroad in years, duration of post doctoral study in India in years and the Volume of communication in terms of average number of email messages received and sent by a scientist in a week.

1. The simple correlation between the group size and age has the negative correlation. It shows that there is no relationship between these two variables.
2. In the case of the correlation between the number of collaborations with age and group size also shows negative association.
3. The volume of communication has a l positive correlation with number of collaborations and duration of postdoctoral study abroad but it has negative association with age, duration, group size and postdoctoral study in India.
4. PDFIDU has positive correlation with all the other explanatory variables except PDFAD. PDFADU has negative association with all other variables.
5. The number of collaborations has positive correlation with DURATION.

Table.7.2: Correlation Matrix

	AGE	DURATION	GPSIZE	NOCOLAB	PDFADU	PDFIDU	VOLCOM
AGE	1.000						
DURATION	.8139 p=0.000	1.000					
GPSIZE	-0.1029 p=0.468	0.0412 p=0.772	1.000				
NOCOLAB	-0.0373 p=0.793	0.0046 p=0.974	-0.2486 p=0.076	1.000			
PDFADU	-0.1050 p=0.459	-0.3610 p=0.009	-0.255 p=0.858	-0.816 p=0.565	1.000		
PDFIDU	0.0982 p=0.489	0.2235 p=0.111	0.2762 p=0.047	0.0294 p=0.836	-0.2719 p=0.051	1.000	
VOLCOM	-0.1671 p=0.236	-0.2721 p=0.051	-0.1103 p=0.436	0.3764 p=0.006	0.3746 p=0.006	-0.1123 p=0.428	1.000

7.3 Regression Analysis

Regression analysis is a branch of statistical theory that is widely used in all most all the scientific disciplines. It is the basic technique for measuring or estimating the relationship among the variables in the study.

Table.7.3: Results of the Multiple Regression Analysis

S L.NO	DEPENDENT VARIABLE	INDEPENDENT VARIABLES	ESTIMATES	STANDARD ERROR
1.	PRODUCTIVITY	CONTSTANT	-16.704	28.967
2.		AGE	0.371	0.848
3.		DURATION	0.515	0.860
4.		GPSIZE	1.317	0.480
5.		NOCOLLAB	4.791	1.284
6.		PDFADU	0.433	1.736
7.		PDFIDU	-6.251	3.082
8.		VOLCOM	0.0580	0.063

R^2=0.511
Adjusted R^2=0.435

The estimated multiple regression coefficients for the productivity among organic chemists is shown in the Table No. 7.4. The value of R Square is found to be 0.511 and adjusted R Square, 0.435 and the Standard Error of the estimate is 18.0598. The model accounts for 51.1 per cent of the variability in the productivity among Organic Chemists, with the other independent variables such as Age, Duration, Group Size, Number of Collaborations, Duration of Post Doctoral Study Abroad, Duration of Postdoctoral Study in India and the Volume of Communication.

Multiple regression analysis suggests that,

1. One Unit increase in the Age will lead to increase in productivity of organic chemists by 0.371 units.
2. The unit increase in the average size of the research group is directly associated with 1.317 units of productivity of organic chemists.

3. The duration of the scientists' association with organization is associated with the productivity of organic chemists by 0.515 units.
4. As compared with the other variables, the number of collaborations has more significant effect on the productivity of Organic Chemists. The result shows increase in one unit in the number of collaborations will lead to increase in 4.791 units of productivity of organic chemists.
5. The Volume of communication in terms of the average number of e-mail messages sent and received by scientists in one week has a low positive impact on the productivity of the scientists. One- unit increase in the volume of communication will increase the productivity of scientists by 0.0580 units.
6. The impact of the Duration of Post Doctoral Study abroad has been identified to have a positive impact on the productivity of organic chemists by 0.433 units., i.e., for every unit increase in post doctoral training abroad change in productivity is 0.433 units.
7. Whereas the impact of the Duration of Post Doctoral Study in India has a negative relationship with productivity of organic chemists. The result shows one unit increase in the Duration of Post Doctoral Study in India will lead to a decrease of 6.251 units in productivity of organic chemists.

CHAPTER EIGHT

CONCLUSION

Sociologists have been engaged in describing and explaining systems of stratification in societies. Sociologists of science have been attempting to understand the stratification in science by focusing on factors responsible for social inequality in the institution of science. In our study we aimed at understanding stratification in Indian science, through a study of productivity differentials among members of the community organic chemists in different organizational contexts.

Production of scientific knowledge is a social activity carried out in the scientific organizations by scientists. Robert Merton (1973) a functionalist argues that production of knowledge takes place within a framework of norms, which he calls ethos of science. Price (1963) explained the growth of scientific knowledge as an exponential function with logistic model of growth. Thomas Kuhn (1962) proposed that production of knowledge in science is a revolutionary process of continuous cognitive consensus (closed model of science) and social constructivists like Karin Knorr-cetina (1978) advocated the role of internal and external socio-cultural contexts in the social production of scientific knowledge. Allison and Scott (1982) considered organizational context to explain the production of knowledge measured as productivity.

It is considered that scientific organizations possess economic, cultural and human resources that vary within and across different socio-cultural contexts. Organizations shape trajectory of scientists, by providing training and facilitating social processes related to production of knowledge like communication and collaboration. Thus we hypothesized that, structural factors like doctoral affiliation, postdoctoral training, area of specialization in which a scientist is engaged, average size of the scientists' research group, volume of communication and number of collaborative projects seem to be responsible, for the variation of productivity among the organic chemists.

Production of knowledge is influenced by the normative structure of science. In our study we started with the idea of changing normative structures of science from the Mertonian ethos (1973) to those of the Mode2 production of knowledge (Gibbons *et.al*: 1994). We observed that there is a disproportionate concentration of resources among organic chemists, in some of the elite organizations thus indicating inequality. We also observed from the data obtained that there is a noticeable tendency towards inter-disciplinarity especially among the organic chemists in the age group of 40-49 years.

Internationalization of science is noticed in the sense that majority of the scientists in the study are in collaborative research with scientists abroad. About 90% of the scientists from elite scientific institutions in India like the deemed university, central university, the three IITs, and the national laboratories included in our study are involved in collaborative research with scientists from abroad compared to a mere 50% from the regional universities.

Russel and Galina (1998) advocated the urgency of building indigenous scientific capacity as they found that scientific communities in the developing countries are isolated from mainstream science. In the Indian context, among organic chemists, we observed that, the scientists who are affiliated to elite institutions have more international linkages than those who are located in regional universities do. Scientists affiliated to regional universities are restricted to other regional institutions in India and their linkages with scientists in elite institutions in India are weak. It appears that the organic chemists from provincial institutions tend to construct invisible college of their own parallel to the invisible colleges comprised of scientists from elite institutions who are a part of the international scientific community and invisible colleges in their research specialties. Thus, communication structure in science in India seems to be a two layered non-intersecting, parallel.

Schurm and Morris (2000) in their work identified that international and domestic networks are inversely related. From our study we found that, while the networks, especially those with scientists abroad are mainly based on professional considerations, the networks that scientists from regional institutions tend to be in, are individual-centered and personality based.

The study attempted to find out the factors that influence productivity among the organic chemists in India. Ben-David (1971) observed that social context of science is directly related to scientific productivity. Diana Crane (1979) argued that productivity depends on the stage of the cognitive development of the discipline. In our study we observed that organic chemistry started as a traditional academic discourse, and now emerged as the biggest inter-disciplinary science, with frontier areas of bio-chemistry, and supra-molecular chemistry. We observed that, majority of the scientists from the elite institutions in the relatively younger age groups like those between 40-49 years more so, 40-44 years are engaged in research in the frontier areas and are highly productive. Frontier areas provide more opportunities for producing original research, which in enhances the recognition level of scientists. The majority of scientists engaged in traditional areas of research in organic chemistry are mostly concentrated in regional institutions and their productivity levels are lower compared to those who are engaged in frontier areas.

In the study I observed institutional inequality in terms of doctoral affiliation and recruitment. I identified a two-layered stratification in Indian Science, the upper layer, comprising of the elite institutions funded by federal central government and the lower, to the regional institutions that are funded by the state/regional governments. I observed that those who obtained doctoral degree from elite institutions ranked high in productivity and all the scientists who obtained doctoral training from regional institutions were low in their productivity level. Only those trained in the elite institutions both in India and abroad are recruited in elite institutions. Social factors external to science like state policies, influence of political power play a significant role among the regional organizations. A tendency to recruit Ph.Ds trained at the same regional university has become conspicuous contributing to high levels of inbreeding in the regional universities. As a result no qualified person from the elite institutions gets recruited by the regional institutions.

Elite institutions emphasize upon, as a norm, a minimum period of postdoctoral training for first appointment. We observed that doctorates from the elite institutions move to scientific organizations abroad for cognitive needs and exposure to better technology in research. Whereas, regional institutions emphasize upon the ability to teach and in the prolonged absence of recruitment, due to constraints placed by the regional governments. Doctorates from state universities migrate abroad for better salaries, and alternative employment in industry. The majority of the scientists in the age group 40-49 employed in elite institutions gained postdoctoral experience in reputed foreign universities before their first regular appointment. The productivity of those who gained postdoctoral experience of 2 to 2.5 years is higher than those whose postdoctoral experience is either less than 2 years or more than 2.5 years. The post doctoral experience of 2-2.5 years seems to have greater influence on productivity than either below 2 or above 2.5 years. We may infer that 2-2.5 years is the optimal duration of postdoctoral training, which shall have positive effect on later productivity levels.

It is observed that economic resources are generally scarce. Organic chemists, especially those located in elite institutions receive only up to 25% of the expenditure towards the research as in-house support. To augment the resources scientists mobilize funds from different public and private funding bodies.

It appears that scientists who are pursuing research in frontier areas that cut across disciplines seem to be successful in mobilizing funds from public funding bodies like Department of Atomic Energy (DAE) and Department of Biotechnology (DBT) besides University Grants Commission (UGC) and Council for Scientific and Industrial Research (CSIR). This trend may be seen as mode 2 production in Indian context where the national funding bodies are recognizing significance of interdisciplinary research also.

In the study, I attempted to explore the development of the areas of specialization wherein we observed that scientists in the relatively younger age group (40-49 years) in elite institutions specialize in the frontier areas of research and those from the regional institutions from all age groups specialize in the traditional areas of research. Those of the organic chemists in the frontier areas of research are engaged in highly inter-disciplinary areas of research and are high in productivity levels. Those in the traditional areas of research are low in productivity levels. As mentioned earlier, new areas of research such as supra molecular chemistry and biomolecules provide more opportunities for original research.

We examined the information management system among the scientific organizations. We observed that CMC has been reshaping the information management system in scientific organizations. As a result of laboratory-library nexus a new techno-social structure has come into being. The elite institutions in our study are into a digital world of research as they have access to instruments of communication technology. The advent of communication technology and its deployment in Indian science is about a decade old.

The study revealed that the majority organizations do not provide the scientists with communication technology as part of the in-house support. Organic chemists acquired instruments of communication technology from research grants. Scientists from the elite institutions use electronic communication more extensively and are more productive compared to those who are located in regional institutions. Scientists from elite institutions have direct access to and control over information and communication technology. They have these instruments right on their desks and in labs. They use this mode of communication for establishing productive links with peers located in India and abroad. In case of the regional institutions, the techno-social structure of library - laboratory has yet to emerge. Libraries and laboratories suffer from the absence of quality journals and absence of computer mediated communication technology that links library and laboratory. We observed that scientists from regional institutions seem to have limited access or no access to electronic communication and so the extent of their usage is limited. It implies that they are not in a position to establish productive linkages with the other organic chemists in elite institutions in India and their counterparts abroad.

In the study I found that the use of electronic mode of communication facilitated faster and economic means of communication that facilitates production of knowledge through collaboration and communication of research output to peer groups and journals. We observed that the communicative needs of the scientists are very high in the age group of 40-49 and the volume of electronic communication is positively correlated with high productivity levels. We conclude that those who have access to CMC tend to be highly productive and those who do not lag behind. Thus the digital divide in the context of scientific research in India is likely to reinforce the social inequality. This stratification is dysfunctional to the organic chemists in the regional universities.

I observed the prevalence of two trends of collaboration in the Indian context. The first category of collaboration is with scientists abroad and the second, collaboration with scientists within India. The data reveals that collaboration of the first category is based on shared cognitive interests whereas collaboration of the second category is limited to personalities.

Collaborative efforts of organic chemists in elite institutions with their counterparts in foreign countries is based on needs related to theoretical, experimental and physical resources like instruments and funds. As mentioned earlier, the weak linkages between organic chemists from the regional institutions on the one hand and scientists in the elite institutions on the other are driven by the urge to catch up with the latter.

The first category of collaboration seems to promote democratic distribution of credit based on respective contribution reflected by the pattern of authorship in the publications. The second category seems to lead to problems in credit-sharing resulting in hierarchy in the pattern of authorship. We observed that collaborative efforts are very high in the age group of 40-49 years and the productivity of the scientists seems to depend on the number of collaborative projects.

The study revealed that in the Indian context, productivity of scientists is viewed still in quantitative terms as the number of publications in the refereed journals is seen as an index of productivity. Consideration of indices of quality like the journal impact factor and citation index that are generally treated as proxy to quality of publications are not yet institutionalized in a significant way.

The findings presented above, are also supported by statistical analysis of the relation between: productivity (dependent variable) and age, duration of employment in the current organization, duration of postdoctoral training abroad, duration of postdoctoral training in India, average size of the research group, volume of electronic communication and number of collaborative projects (independent variables).

Multiple regression analysis suggests that,

1. One Unit increase in the Age will lead to increase in productivity of organic chemists by 0.371 units.
2. The unit increase in the average size of the research group is directly associated with 1.317 units of productivity of organic chemists.
3. The duration of the scientists' association with organization is associated with the productivity of organic chemists by 0.515 units.
4. As compared with the other variables, the number of collaborations has more significant effect on the productivity of Organic Chemists. The result shows increase in one unit in the number of collaborations will lead to increase in 4.791 units of productivity of organic chemists.
5. The Volume of communication in terms of the average number of e-mail messages sent and received by scientists in one week has a low positive impact on the productivity of the scientists. One- unit increase in the volume of communication will increase the productivity of scientists by 0.0580 units.
6. The impact of the Duration of Post Doctoral Study abroad has been identified to have a positive impact on the productivity of organic chemists by 0.433 units., i.e., for every unit increase in post doctoral training abroad change in productivity is 0.433 units.
7. Whereas the impact of the Duration of Post Doctoral Study in India has a negative relationship with productivity of organic chemists. The result shows one unit increase in the Duration of Post Doctoral Study in India will lead to a decrease of 6.251 unit in productivity of organic chemists.

Suggestions for further Studies

1. Given the variety in organizational contextually and differences in the evaluation patterns of scientific productivity in India, there is a

necessity for further studies that will lead to suggestions regarding indices of quality of productivity.

2. Socio-economic progress of a developing country like India relies upon scientific progress. In this context, we suggest that the policy-making bodies of the government encourage collaboration in inter-disciplinary research among the scientific organization. This can be achieved by
    a. Increasing funding for inter-disciplinary programs/research
    b. Introducing inter-disciplinary courses in all sciences, in universities at masters level and
    c. Organizing conferences/seminars/workshops on inter-disciplinary themes more often.
3. The academic institutions must recognize the importance of postdoctoral training and must institutionalize high quality Postdoctoral Training programs and associated infrastructure within the country.
4. The Government of India must realize the importance of communication technology in scientific research and incorporate the instruments related to CMC as a one-time effort/release separate funds, in addition to the in-house budgetary support to the scientists.
5. The scientific bodies must increase financial support for collaborative exchange programs in frontier areas of organic chemistry.

Limitations

This work is a modest attempt to understand the determinants of productivity and dynamics of interaction among community of organic chemists in India. Further studies in other disciplines are required to understand the dynamics of science in India fully. Such studies also should focus on how to minimize stratification in Indian Science.

BIBLIOGRAPHY

Adams, Johnson and Wilcox C. 1970. *Laboratory Experiments in Organic Chemistry*. London: The Macmillen Company Ltd.

Barnes, B. 1974. *Scientific Knowledge and Sociological Theory*. London: Rouletdge and Kagan Paul.

Beaver, D. De B. and R. Rosen. 1979. Studies in scientific collaboration Part II-. Scientific co-authorship, research productivity, and visibility in the French scientific elite, *Scientometrics,* Vol.1, pp.133-449

Beaver,D & R.Rosen .1979b.'Studies in Scientific Collaboration: Part III-Professionalization and the Natural History of Modern Scientific co-authorship', *Scientometrics* 1(3), pp.133-49.

Ben-David. 1971. *The Scientist's Role in Society*. Englewoood Cliffs. New Jersey: Prentice-Hall.

Bernard, Miller. 1971. *Organic Chemistry, the Basis of life*. Massachusetts: The Benjamin Cummiungs Publishing Company Inc.

Blau, J. 1974. 'Patterns of Communication Among High-Energy Physicists', *Sociometry*, Vol. 37(3): 391-406.

Blekhodja.O and R. Landry Triple Helix collaboration?why do researchers collaborate with industry and government?What are the factors influencing the perceived barriers? In 5th Triple Helix Conference, Turin, Italy.(2005).

Bordwell, F.G. 1963. *Organic Chemistry*. New York: The Macmillen Company.

Bozeman, B and E Corley.2004. 'Scientists' collaboration strategies: implications for scientific and technical human capital', *Research Policy*, 33, pp.599-616.

Bruce, V Leuvenstein. 1995. 'From Fax to Facts: Communication in the Cold Fusion Saga', *Social Studies of Science,* Vol. 25: 403-36.

Burgoyne, Edward E. 1979. *A Short Course in Organic Chemistry*. New York: The McGraw-Hill Book Company.

Cambrosio, Alberto, Camile Limoges, and Densye Pronovost. 1991. 'Analyzing science Policy Making: Political Ontology of ethnography', Social Studies of Science, Vol. 4: 775-81.

Carey, F. A. 1986. Organic Chemistry. New York: The McGraw-Hill Book Company.

Caroline S.Wagner.2005. 'Six case studies of international collaboration in science', *Scientometrics*, 62 (1) pp.3-26.

Clarke, Adele E., and Joan H. Fujimura (ed.). 1992. *The right tools for the job: At work in twentieth-century life sciences.* Princeton: NJ: Princeton University Press.

Cole,S.1979. 'Age and Scientific Performance', *American Journal of Sociology* 84(4):958-977

Cram, and Hammond G. 1964. Organic Chemistry (2nd Ed.). New York: The McGraw-Hill Book Company.

Crane, Diana. 1972. *Invisible Colleges: Diffusion of Knowledge in Scientific Communities*. Chicago and London: The University of Chicago Press.

Crow, Micheal and Barry Bozeman.1987. 'A new typology for R&D laboratories: Implications for policy analysis', *Journal of Policy Analysis and Management*, Vol. 6: 328-41.

Donald, C. Pelz & Frank M. Andrews. 1966. *Scientists in Organizations, Productive Climates for Research and Development*. New York: John Whiley and Sons, Inc.

Douglas, Leckber & Glester Hill Jr. 1979. 'The Paradigm Concept and Sociology: A Critical Review', *American Sociological Review*, Vol. 44: 925-937.

Drucker, P. 1988. 'The Coming of the New Organization', *Harvard Business Review*, Vol. 45-53.

Edward J Hacket(2005), Introduction to Scientific Collaboration', *Social Studies of Science*, 35(5),pp. 667-671.

Elias, James Corey. 1990. 'The logic of Chemical Synthesis: Multistep Synthesis of complex Carbogenic Molecules, Nobel Lectures in Chemistry 1981-1990', in Tore Fransmyr, Bog Malimstrom (ed.). *World Scientific*, Singapore: Nobel foundation.

Ferguson, Lloyd N. 1963. *The Modern Structural Theory of Organic Chemistry*. London: Prentice-Hall International Inc.

Gabriel, Weimann. 1982. 'On the importance of Marginality: One more step into the two -step flow of Communication', *American Sociological Review*, Vol. 47: 764-773.

Galison, P&Hevly, B.1992. (Eds) Big Science: the growth of large scale research. Stanford,CA: Stanford University Press,

Garvey, W.D. & Griffith, B.C. 1968. 'Informal Channels of Communication in the Behavioural Sciences: Their Relevance in the Structuring of Formal Bibliographic Communication 129-146', in Montgomery, E.B. (ed.). *Foundations of access to knowledge.* Syracuse: NY: Syracuse University.

Garvey, W.D. 1979. *Communication: The Essence of Science*. Oxford: Pergamon.

Gaston, Jerry, 1970 'The Reward System in British and American Science', *American Sociological Review* Vol. 35: 718-732

Gerhard, Sonnert. 1995. 'What makes a Good Scientist?: Determinants of Peer Evaluation among Biologists', *Social Studies of Science,* Vol. 25: 35-55.

Gibbons, M, Limoges, C, Nowotny, H, Schwartzman, S, Scott, P and Trow, M. 1994. '*The New Production of Knowledge: The Dynamics of Science and Research in Contemporary Societies*. London: Sage

Gibbons, Micheal, Camille Limoges, Helga Nowotny, Simon Schwarzman, Peter Scott, and Martin Trow. 1994. *The Production of Knowledge*. London: Sage.

Godin, B. 1998. 'Writing performative history: the new Atlantis?', *Social Studies* of *Science*, Vol. 28 (3): 465–83.

Godin, Benoit & Yves Gingras (2000), 'Impact of Collaborative Research on Academic Science', *Science and Public Policy* 27(1), pp. 65-73.

Griffin, Rodger W. 1969. *Modern Organic Chemistry*. New York: The McGraw-Hill Book Company.

Habermas, .J. 1984 *The Theory of Communicative Action*, Vol.1& 2 Cambridge, U. K. Polity Press.

Hagstrom .W 1964. Traditional and modern forms of scientific team work, *Administrative Science Quarterly*, 9 (3), pp. 241-264.

Hammett, Louis P. 1970. *Physical Organic Chemistry*. New York: The McGraw-Hill Book Company.

Haribabu, E. 1991. 'A Large Community of Scientists but few peers: A study of the Scientific Community in India', *Sociological Bulletin*, Vol. 40 (1&2): 217-233.

—. 1999. 'Scientific Knowledge in India: From Public Resource to Intellectual Property', *Sociological Bulletin,* Vol. 48 (1&2): 217-32.

Helga Bermeo Andrade et al. 2009.'Dimensions of scientific collaboration and its contribution to the academic research groups' scientific quality', *Research Evaluation*, 18 (4), 301-311.

Hendrikson, Cram, and Hammond G. 1970. *Organic Chemistry* (3rd Ed.). New York: The McGraw-Hill Book Company.

Henry, Ezkowitz and Loet Leydesdorff.2000.The dynamics of innovation', *Research Policy*, Vol. 29 (2), pp.109.

Hillary, Rose and Rose. 1977. *Science and Society*. Harmondsworth, Midlesex, England: Penguin Books Ltd.

Ignatyev, A. A. & Mirskaja, E.Z. 1989. 'The Problem of Going from: Science Policy and Human Factors in the Experience of Developing Countries', Social Epistemology, Vol.3 (3): 217-227.

Ingold, C.K. 1969. Structure and Mechanism in Organic Chemistry. Ithaca: Cornell University Press.

J Sylvan Katz & Ben R. Martin .1997. 'What is research collaboration?' *Research Policy* 26, pp.1-18

Jacobs, Struan. 1987. 'Scientific Community: Formulations and Critique of a Sociological Motif', *The British Journal of Sociology*, Vol. 38 (2): 266-276.

Jacques, Gaillard, V. V. Krishna and Rowland Waast. 1997. *Scientific Communities in the Developing World*. New Delhi: Sage Publications.

Jacques, Gaillard. 1991. *Scientists in the Third World.* Kentucky: University of Kentucky Press.

Jairath, Vinod K. 1984. 'In Search of Roots – The Indian Scientific Communication', *Contributions to Indian Sociology*, Vol. 18 (1): 109-130.

James, B. Hendrickson, Donard J. Cram, George S Hammond. 1970. *Organic Chemistry*. New Delhi: McGraw-Hill, KogaKusha, Ltd.

James, E Curtis & John. W. Petras. 1970. *The Sociology of Knowledge: A Reader*. Praeger Publishers Inc.

James, Robert Brown. 1984. *Scientific Rationality: The Sociological Turn.* Boston: D. Reidel Publishing Company.

James, W. Endersby. 1996. 'Collaborative Research in the Social Sciences: Multiple Authorship and Publication Credit', *Social Science Quarterly*, Vol. 77 (2).

Johnston, R. 1994. 'Effects of resource concentration on research performance'. *Higher Education,* 28(1), 25-37

Jonathan, R. Cole & Stephen Cole. 1973. *Social Stratification in Science*. Chicago: The University of Chicago Press.

Jordan, J.M., Meador,M & Walters,S.1988.Effects of department size organization on the research productivity of academic economists. *Economics of Education Review,* 7(2),251-255.

Joseph, Agassi. 1981. *Science and Society, Studies in the Sociology of Science*. London: D. Reidel Publishing Company.

Jurgen Enders.2005 'Research Training, Knowledge Dissemination and the Transformation of Academic Work, '*Higher Education*, Vol.49,No.1/2:119-133.

Karen Seashore , Janet M.Holdsworth,Melissa S.Anderson and Eric G.Campbell. 2007. 'Becoming a Scientist: The Effects of Work-Group

Size and Organizational Climate',*The Journal of Higher Education,* Vol.78 No.3,311-336

Katz, J.S. and Martin B.R. 1997. 'What is research Collaboration?', *Research Policy*, Vol. 26: 8-9

Keith, J. Laider. 1995. 'Lessons from the History of Chemistry', *Acc.Chem.Res*, Vol. 28: 187-192.

Kircz, Joost G. 1998. 'Modularity: The Next Form of Scientific Information Presentation', *Journal of Documentation*, Vol. 54 (2): 210-235.

Kling, Rob. 1980. 'Social Analysis of Computing: Theoretical Perspectives in Recent Empirical Research', *Computing Surveys,* Vol. 12 (1): 61-110.

Kloppenburg, Jack R. Jr. 1988. *First the seed: The political economy of plant biotechnology, 1492-2000.* New York: Cambridge University Press.

Knorr, Karin, Mittermeir, Roland, Aichholzer, Georg, Waller, Georg. 1978. 'Individual Publication Productivity as a Social Position Effect in Academic and Industrial Research Units', in Andrews Frnak (ed.) *Scientific Productivity: the Effectiveness of Research Groups in Six Countries.* London: Cambridge University Press.

Knorr-Cetina & Mulkay Micheal. 1983. *Science Observed.* London: Sage Publications.

Knorr-Cetina, Karin D. 1981. *The manufacture of knowledge: An essay on the constructivist and contextual nature of science.* Oxford: Pergamon press.

Knorr-Cetina, Karin. 1995. 'How Superorganisms change: Consensus formation and the social ontology of high energy-physics experiments', *Social Studies of Science,* Vol. 25: 119-47.

—. 1998. *Epistemic cultures: The Cultures of Knowledge Societies.* Cambridge: Harvard University Press.

Krishna, V.V. 1991. 'The Emergence of the Indian Scientific Community', *Sociological Bulletin,* Vol. 40.

Krohn, R.G. 1971. *The social shaping of Science.* West port, Conn: Greenwood Publishing.

Kronick, D.A. 1988. 'Anonymity and Identity: Editorial Policy in the Early Scientific Journal Library', *Quarterly*, Vol. 58: 221-229.

Kuhn, Thomas. 1962 (1st ed.). *The Structure of Scientific Revolutions.* Chicago: The University of Chicago Press.

Lacetera,N.,Cockburn,I.M.Henderson,R 2004. Do firms change capabilities by hiring new people? A study of the adoption of science-based drug discovery. In J.A.C.Baum&A.M.McGahan(Eds.),Business strategy over the industry life cycle,Vol.21,pp.133-159.

Latour, Bruno. 1983. 'Give me a laboratory and I will raise the world 141-70, in Karin D. Knorr-Cetina and Micheal Mulkay (ed.). *Science Observed.* New Delhi: Sage.

—. 1987. *Science in action: How to follow scientists and engineers through society.* Cambridge: Harvard University Press.

Leah, A. Lievrouw & Kathleen Carley. 1990. 'Changing patterns of Communication Among Scientists In an Era of TeleScience', *Technology in Society,* Vol. 12: 457-77.

Lehman, H.C.1953. Age and Achievement, Princeton, N.J.;Princeton University Press

Leslie, Stuart W. 1993. *The cold war and American science: The military-industrial-academic complex at MIT and Stanford.* New York: Columbia University Press.

Linus Pauling, 1960. *The Nature of the chemical Bond and the structure of molecules and crystals: An Introduction to Modern Structural Chemistry*: New Delhi: Oxford & IBH Publishing Co.

Long, J. Scott and Robert Mc Ginnis. 1981. 'Organizational context and Scientific Productivity', *American Sociological Review*, Vol. 46: 422-442.

Lotka, Alfred J 1926.'The Frequency Distribution of Scientific Productivity', *Journal of the Washington Academy of Science*, 16, pp.317-23.

—. 1926. 'The Frequency Distribution of Scientific Productivity', *Journal of the Washington Academy of Science* 16:317-23

Lowell, Hargens & Diane, H Felmlee. 1984. 'Structural Determinants of Stratification in Science', *American Sociological Review*, Vol. 49: 685-697.

Maienscbein, Jane. 1993. 'Why collaborate?' *Journal of History of Biology,* Vol. 26: 167-83.

Maresi Nerad and Joseph Cerny 1999 'Postdoctoral Patters, Career Advancement, and Problems' *Science,* New Series, Vol.285, and No.5433:1533-35

Markus, Lynne and Daniel Robey. 1988. 'Information Technology and Organizational Change: Causal Structure in Theory and Research', *Management Science*, Vol. 34 (5): 583-98.

Mary, Frank Fox.1983. 'Publication Productivity among Scientists: A Critical Review', *Social Studies of Science,* Vol 13, No.2 pp.285-305

Mats, Benner. 2000. 'Advantages and disadvantages of scientific networking in the era of 'globalization', Revised Version 011029

—. 2000. 'Advantages and disadvantages of scientific networking in the era of 'globalization', Revised Version 011029

Maurice, Crosland. 1990. 'The History of Chemistry Seen in a Broader Context', *Impact of Science on Society*, Vol. 159: 227-236.

McCarthy, Thomas, A. 1979. *A Critical Theory of Jurgen Habermas*, Cambridge, MIT Press

Melin, Goran.2000. 'Pragmatism and Self-Organization: Research Collaboration on the Individual Level', *Research Policy* 29(1), pp. 31-40

Menzel, H. 1966. 'Scientific Communication: Five Items From Social Science Research', *American Psychologist*, Vol. 21(11): 999-1004.

Merton, R.K. 1957. *Social Theory and Social Structure* (revised ed.). New York: Glencoe Free Press.

—. 1973a. 'The Normative Structure of Science 267-78', in R.K. Merton (ed.). *Sociology of Science.* Chicago: University of Chicago Press.

—. 1973b. 'Priorities in Scientific Discovery 286-323', in R.K. Merton (ed.). *Sociology of Science.* Chicago: University of Chicago Press.

—. 1973c. 'Age, Aging, and Age Structure in Science', in R.K. Merton (ed.). *Sociology of Science.* Chicago: University of Chicago Press

Micheal Mulkay.1979. *Science and Sociology of Knowledge.* London: George Allen & Unwin Ltd.

—. 1980. 'The Sociology of Science in East and West', *Current Sociology*, Vol. 28 (3): 22-116.

—. 1980. 'The Sociology of Science in East and West', *Current Sociology*, Vol. 28 (3): 22-116.

Mikhailow, A.I & chernyi, A. & Giliarevskii, R. S. 1984. *Scientific Communications and informatics.* Arlington: Information Resources Press.

Misa, Thomas J., 1992. 'Theories of Technological Change: Parameters and purposes', *Science, Technology and Human Values*, Vol. 17 (1).

Moore, Kelly. 1995. "When inside and outside matter Scientists, Activism and Changes in Scientific Knowledge". Paper presented at the annual meeting of the, October. *Society for Social Studies of Science*

Moravscik, M.J. 1978. 'Do less developed countries have a special science of their own?', *Interceincia*, Vol. 3: 8-13.

—. 1985b. 'Science In The Developing Countries: An Unexplored Fruitful Area for Research in Science Studies', *Society for Social Studies of Science Review*, Vol. 3: 2-13.

Morgan, Gareth. 1986. *Images of Organization.* Beverly Hills: Sage

Muniyappa.K.2007."The role of postdoctoral training for careers in research and higher education ', *Current Science* Vol.92.No.4:450-454

National Research Council. 1976. *Survey of Biomedical and Behavioral Scientists*. Washington: GPO.

Nebergall, W.H., Schmidt, Frederic, Holtzclaw, and Henry. 1980. *Chemistry* (6th Ed.). London: Pantheon.

Ngwenyama, O, Orlikowski, W and Barouldi, I. 1991. 'Studying Information Technology in Organizations: Research Approaches and Assumptions', *Information Systems Research*, Vol. 2 (1): 1-28.

Noble, David. 1984. *Forces of production: A social history of Industrial Automation*. New York: Knopf.

Noriko Hara et al 2003. 'An Emerging view of Scientific Collaboration: Scientists' Perspectives on Collaboration and Factors that Impact Collaboration', *Journal of the American Society for Information Science and Technology* 54 (10), pp.952-965.

Norman W.Storer.1966. The Social System of Science.NewYork; Holt, Rineheart and Winstone

Orbuch, T. 1997. 'People's accounts count: The sociology of accounts', *Annual Review of Sociology*, Vol. 23: 455-78.

Owen-Smith, Jason D (2005), 'Dockets, Deals and Sagas: Commensuration and the Rationalization of Experience in University Licensing', *Social Studies of Science* 35(1), pp. 69-97.

Pelz, D.C. and F.M.Andrews .1966. *Scientists in Organizations. Productive Climates for Research and Development.* New York: John Wiley and Sons

Porter E.Coggeshall, John C.Norvell, Lawrence Bogorad, Robert M Bock 1978 'Changing postdoctoral Career Patterns for Biomedical Scientists' *Science,* New Series, Vol.202,No.4367:487-493

Price, Derek J.de Solla &Donald Beaver 1966. 'Collaboration in an Invisible College', *American Psychologist, 21*, pp.1011-18.

Pugh, Derek, D. Hickson, and C.Hinings. 1969. 'An empirical taxonomy of work organizations', *Administrative Science Quarterly*, Vol. 14: 115-26.

Rahman, A. 1975. 'Goals for basic and applied research in different national and cultural contexts', Journal of Scientific and Industrial Research, Vol. 34: 1-7.

Ricardo B. Duque (et.al) 2005. 'Collaboration Paradox: Scientific Productivity, the Internet, and Problems of Research in Developing Areas', *Social Studies of Science*, 35(5), pp.755-785

Richard, Whitley. 1984. 'The Modern Sciences as reputational work Orgnaizations 1-41', in Richard Whitley (ed.) *The Intellectual And Social Organization Of The Sciences*. Oxford: Clarendon Press.

Robert, Mc Ginnirs, Paul D. Allison and J. Scott Long. 1982. 'Postdoctoral Training in Bioscience: Allocation and Outcomes', *Social Forces*, The University of North Carolina Press. 701-722.

Rogers, Everett M. 1993. 'The Diffusion of Innovations Model 9-24', in I. Masser and H.J. Onsurd (ed.). *Diffusion and Use of Geographic Information Technologies*. Dordrecht: Kluwer.

Russel, J.M & Galina, C.S. 1998. 'Basic and Applied Research in Developing Countries: The search for an Evaluation Strategy', *Knowledge & Policy*, Vol. 10: 102-114.

Schild, I. 1997. "International collaboration as work in polar research". Paper presented at the annual meeting of the *Society for Social Studies of Science*, Tucson: Arizona.

—. (1997) "International collaboration as work in polar research". Paper presented at the annual meeting of the Society for Social Studies of Science,.Tucson: Arizona.

Schrum, Wesley, and Joan Morris. 1990. 'Organizational constructs for the assembly of technological knowledge 235-59', in S. Cozzens and T. Gieryn (ed.). *Theories of science in society.* Bloomington: Indiana University Press.

Shapin, Steven. 1988. 'Following Scientists around', *Social Studies of Science*, Vol. 18: 533-50.

Sharon G.Levin&Paula.E.Stephan.1988.Age and research productivity of Academic Scientists: *Research in Higher Education,* Vol.30 No.5,pp 531-549

Shrum, Wesley and Patricia Campion. 2000. 'Are Scientists in Developing Countries Isolated', *Science,Technology and Society*, Vol. 5(1).

—. 1998. 'Review Essay: The Labyrinth of Science', *American Journal of Sociology*, Vol. 94 (2): 396-403

Singer, Maxine. 2000. 'Enhancing the Postdoctoral Experience', *Issues in Science and Technology*, Vol. 17 (1): 43-44

Smit, W. A., A.F. Bochkov and R. Caple. 1998. *Organic Synthesis: The Science behind the Art.* Cambridge, UK: The Royal Society of Chemistry.

Solla Price, Derek. J.de 1986. *Little Science, big science and beyond,* Newyork

—. 1963. Little Science, Big Science. New York: Columbia University Press.

Sooho Lee and Barry Bozeman 2005. 'The Impact of Research Collaboration on Scientific Productivity', *Social Studies of Science,* 35(5) pp. 673-702.

Sproull, Lee S., and Paul Goodman. 1990. "Technology and Organizations: Integration and Opportunities 254-65", in *Technology and Organizations* (ed.). P.S. Goodman, L.S. Sproull and Associates. San Fransico: Jossey-Bass.

Stanley, H. Pine. 1987. *Organic Chemistry* (5th ed.). New Delhi: McGraw-Hill Book Company.

Star, Susan Leigh and James Griesemer. 1989. 'Institutional ecology, translations and boundary objects: Amateurs and professionals Berkeley's Museum of Vertebrate Zoology, 1907-1939', *Social Studies of Science*, Vol. 19: 387-420.

Stewart, F. 1981. 'International Technology Transfer: Issues and Policy Options 67-110', in Streeten P., and R. Jolly (ed.). *Recent Issues in World Development.* Oxford: Pergamon Press.

Stokes, Donald E (1997), Pastuer's Quadrant:Basic Science and Technological Innovation, Washington, DC: Brookings Institution Press

Stronach, I and Maclure 1997. M.. *Educational Research Undone: The Postmodern Embrace*. Buckingham: Open University Press

Traweek, Sharon. 1988. *Beamtimes and lifetimes: The world of high-energy physics*. Cambridge: MA: Harvard University Press.

Turnbull, David, 1997. 'Reframing Science and Other Local Knowledge Traditions', *Futures*, Vol. 29(6): 551-562.

Usher, R. 2000. *All change? The university in the 'real world'.* Keynote paper at the Working Knowledge Conference, 10–13 December 2000. Sydney, Australia: University of Technology.

Walsh, John P. and Todd Bayma. 1996a. 'Computer Networks and scientific work', *Social Studies of Science*, Vol. 26: 661-703.

Walsh, John P. and Todd Bayma. 1996b. *Communication networks and the rise of an information elite.* http://web.mit.edu/marshall/www/home.html.

Weber, Max. 1947. "The Theory of Social and Economic Organization", translated by A.M. Hendersen and T. Parsons. New York: Free Press.

William Zumeata 1984 'Anatomy of the Boom in Postdoctoral Appointments during the 1970s: Troubling implications for Quality Science?' *Science, Technology & Human Values'*, Vol.9, No.2:23-37

Wynne, Brain. 1992. 'Representing policy constructions and interests in SSK', *Social Studies of Science*, Vol. 22: 575-80.

Zabusky, Stacia. 1995. *Launching Europe: An ethnography of European cooperation in space science*. Princeton, NJ: Princeton University Press.

Ziman, John. 1994. *Prometheus bound: Science in a dynamic steady state*. Cambridge, UK: Cambridge University Press.

—. 1996. 'Post-academic Science: Constructing Knowledge with Networks and Norms', *Science Studies*, Vol. 9(1): 67-80.

Zuckerman, H. and R.K. Merton, 1973 'Institutionalized patterns of Evaluation 461-96', in R.K. Merton (ed.). *Sociology of Science*. Chicago: Chicago University Press.